Field Guide to the Weather

Learn to identify clouds and storms, forecast the weather, and stay safe

Ryan Henning

Adventure Publications
Cambridge, Minnesota

Acknowledgments

My gratitude goes out to the brilliant meteorologists who guided me through this process, including Anthony Dunkel, Kelly Faltin, Dr. Kim Klockow-McLain, Brad Nelson, and the countless others whom I have had the privilege of working with or learning from.

A special thank you also goes out to my wife, Katie, and our two boys who were born during the book-writing process, Liam and Easton.

Disclaimer: When it comes to weather safety, always heed official weather warnings, watches and advisories. Do not rely on your own forecasts, or advice from this book, for your own safety.

Cover and book design by Lora Westberg
Photo credits
All photos copyright of their respective photographers.
California Regional Weather Server, sponsored by the Department of Earth & Climate Sciences, San Francisco State University, is licensed under a Creative Commons Attribution 4.0 International License (http://creativecommons.org/licenses/by/4.0; accessed via http://squall.sfsu.edu/crws/jetstream.html): 23 (unaltered); **Copyright by Gerrit Rudolph, Fulda, Germany:** 104 (bottom); **Eric Kurth, NOAA/NWS/ER/WFO/Sacramento:** 114 (top); **NASA:** 14 (bottom); **National Oceanic and Atmospheric Administration/Department of Commerce:** 17; **NOAA National Severe Storms Laboratory:** 125 (top); **NOAA Photo Library, NOAA Central Library; OAR/ERL/National Severe Storms Laboratory (NSSL):** 87 (top); **NOAA:** 10; **NOAA:** 115 (bottom); **NOAA, Greg Carbin:** 87 (bottom); **NOAA/National Weather Service:** 109, 122; **NOAA/NWS:** 117, 124; **NOAA/NWS/National Hurricane Center:** 116; **NOAA/NWS/Storm Prediction Center:** 108, 114 (bottom), 115 (top); **Sean Waugh NOAA/NSSL:** 83, **NWS Kansas City:** 125 (bottom)

Photo credits continued on page 144

10 9 8 7 6 5 4

Field Guide to the Weather
Copyright © 2019 by Ryan Henning
Published by Adventure Publications
An imprint of AdventureKEEN
310 Garfield Street South
Cambridge, Minnesota 55008
(800) 678-7006
www.adventurepublications.net
All rights reserved
Printed in the U.S.A.
ISBN 978-1-59193-824-8 (pbk.); ISBN 978-1-59193-825-5 (ebook)

Table of Contents

Introduction

I grew up in the Twin Cities area, with every facet of weather anyone could be exposed to—severe storms in the summer, blizzards in the winter, and all those sunny days in between. Most meteorologists enter the field because some major weather event piqued an interest that blossomed to a full-fledged passion, but for me, it was the maps that drew me in. I grew up as a "map kid," fascinated by radar, satellite, forecast fronts, and model guidances." Tracking where the weather was headed was endlessly captivating.

I went to Purdue University and got a degree in synoptic meteorology, which deals with predicting large-scale systems (e.g., fronts and predicting high and low pressure systems). I also received a minor in communication, because, in addition to following the weather, I really enjoy the idea of telling everyone about it.

My father worked in the aviation industry, and I followed him there, working in the aviation industry for nearly a decade. During that time, I found that communicating about weather and meteorology had a deeper calling for me, so I started a website, Victoria-Weather.com, named after my hometown of Victoria, Minnesota, in which I could break down the weather and forecasts for anyone who was interested.

The thing about the weather is that it impacts everyone on a daily basis, from your kids' wait for the school bus to the price of your electricity, but, unless you are really interested in meteorology, its terminology and methods aren't intuitive or easy to understand. With my science background and my stated goal from my youth of wanting to tell everyone about the weather, writing this field guide became an immensely appealing idea.

I hope this guide helps you begin to understand meteorology and the weather. If I'm lucky, perhaps you will find something that piques your interest, like it did for me so many years ago.

METEOROLOGY AND
THE BUILDING BLOCKS OF WEATHER

To understand the various phenomena you might see or experience when out in nature, it is important to understand the processes that bring those about. Where do you get rain? From clouds, of course. But how do you get clouds? Atmospheric updrafts are one way. And how are those caused? Sometimes by cold fronts, which are caused by cyclones, which are enhanced by the jet stream, which in turn is a process with roots in the revolution of Earth and its orbit around the sun, dictated by the rules of physics. That seems like a good place to start. In this first section, we'll try to narrow things down so we can really sink our teeth into those phenomena—and how they might affect us every day.

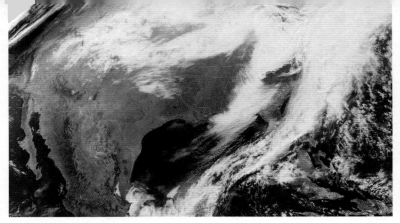

NOAA/GOES EAST, taken January 24, 2019

What Is Meteorology?

Defined simply, meteorology is the study of the processes and phenomena of the atmosphere. To a meteorologist, it is a bit more complicated than that. While most people are interested in the day's forecast, if you want to learn how the weather works, it's helpful to start with a top-down look from the perspective of a professional meteorologist.

At its rawest and most basic level, meteorology is one big math problem. The atmosphere is essentially a big basin full of fluids, such as water vapor and gases (which are considered fluids in physics). These fluids follow all the basic rules of physics, so in theory, the daily forecast is a solvable problem, if it weren't for two issues.

1. Too many calculations are required to solve this equation, and the calculations are too complicated to be solved by the world's fastest computers in time for the forecasts to be usable.

2. There are several equations, called primitive equations, that feed into all of the atmospheric models that meteorologists use for their forecasting, but there aren't enough equations to calculate for the variables in question.

Fortunately, the science of fluid dynamics is well understood, as are the relationships between all the variables, enabling us to process enough of the calculations to get useful forecasts. What's more, because we experience the weather every day and see the results of that big system of fluid dynamics, we understand the phenomena that result from these relationships and can explain how they come about.

That's what meteorology is, in its essence. It's a riddle that is mostly solvable, and any forecast or definition you see—and even this book— is going to be, in its own way, a piece of the overall puzzle.

Planetary Motion

The atmosphere is a whirling mix of gases and water vapor, and it stays active thanks to a combination of the planet's motion, its tilted axis, and its orbit around the sun. Without those factors, the atmosphere would remain sitting still on the Earth's surface, almost like a big lake. Our atmosphere is ever-moving because the Earth itself is constantly in motion; the sun warms a new part of the planet every second.

Earth's Axis and the Seasons

Any study of the weather begins with Earth's orbit and the planet's tilt on its axis. It takes Earth about 365.25 days to orbit the sun; it takes the Earth 24 hours—one day—to rotate on its axis. (The calendar year is rounded down to 365 days, and every four years, we have a leap year, to make up for the difference.)

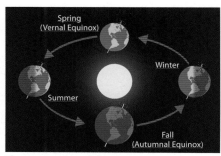

The progression of the seasons

Earth has seasons because it is tilted 23.5 degrees on its axis. As the Earth orbits the sun over the course of a year, this tilt causes some parts of the Earth to get sunlight more directly in parts of the orbit than it does in others. For example, the northern hemisphere gets more direct sunlight in July and the height of summer than it does in February. We refer to this changing relationship with the sun as the changing of the seasons. There are four astronomical seasons in all: Winter, Spring, Summer, and Fall. December 21 marks the day that the Southern Hemisphere is angled most directly at the sun, while June 21 says the same about the Northern Hemisphere. March 21 and September 21 are the days in which night and day are nearly the same length for all latitudes. (These dates are approximate, and can change by about 24 hours in either direction.)

Meteorological Seasons vs. Astronomical Seasons

When it comes to seasons, meteorologists refer to slightly different dates than the astronomical dates listed above. They do so not only for ease of record keeping, but also because they correspond more accurately with when we experience each season's weather. In the Northern Hemisphere, meteorological winter begins on December 1, spring on March 1, summer on June 1, and autumn on September 1.

The Tropics (the area between the Tropics of Cancer and Capricorn) don't see a substantial change of daylight hours from month to month. The Tropics of Cancer and Capricorn are the farthest north and south that the sun can shine directly overhead. Because they see so much direct sunlight every day, there isn't an opportunity for temperatures ever to cool enough in order to welcome the winter.

In general, seasonal variability changes with latitude. The higher the latitude, the more drastic the changes in season. This occurs because there are more drastic changes in direct sunlight.

Changing Amounts of Daylight

The Earth's tilted axis also means that the amount of daylight that a given locale receives varies not only by season, but also by latitude. Areas closest to the poles see the most profound differences in each season, whereas areas close to the equator have far less pronounced differences, thanks to the relatively steady amount of light and heat they receive from the sun. The most extreme example of this occurs within the Arctic and Antarctic Circles. The Arctic and Antarctic Circles are located at 66.5 degrees latitude (north and south, respectively); each circle encompasses 23.5 degrees of latitude—not coincidentally, the Earth's axis is titled at the same angle. It's because of that tilt that the far north and the far south experience polar winters and polar summers.

PLACE	LONGEST DAY (amount of daylight)	SHORTEST DAY
Nome, Alaska	21 hours, 38 minutes	4 hours, 21 minutes
Seattle, Washington	15 hours, 58 minutes	8 hours, 29 minutes
Chicago, Illinois	15 hours, 16 minutes	9 hours, 6 minutes
Washington, DC	14 hours, 53 minutes	9 hours, 26 minutes
Los Angeles, California	14 hours, 26 minutes	9 hours, 53 minutes
Miami, Florida	13 hours, 45 minutes	10 hours, 32 minutes
Honolulu, Hawai'i	13 hours, 26 minutes	10 hours, 50 minutes

During winter in the Arctic Circle, the general area is pointed away from the sun and does not receive as much sunlight. With that said, that doesn't mean it's completely dark for six months; on the contrary, even when the sun is below the horizon, it can still be bright enough to see, even at the poles. This is what is known as polar twilight. A truly dark polar night only occurs for a portion of the polar winter, and only

Arctic twilight in Finland

within a certain distance (above 84 degrees north) of the poles. The same phenomenon occurs in the Antarctic region during its winter.

Polar summer is exactly the opposite; on the summer solstice in the Arctic Circle, the sun never sets, and the same is true in summer in the Antarctic. This is often known as the Midnight Sun. The duration of the Midnight Sun depends on how close one is to the pole.

Differential Heating

While the Earth orbits around the sun, it also spins on its own axis. It may seem intuitive, but the daily change between night and day has a big impact on weather patterns, as it creates differential heating, or an uneven temperature pattern across the planet. Much of our weather is created by the Earth's atmosphere simply trying to reach an equilibrium—a state of balance—and differential heating disrupts that.

What's more, some parts of the planet react differently to increased or decreased access to sunlight. For example, bodies of water, especially large ones, are slower to heat or cool than land masses, and this can create differing interactions between sea and land,

The transition from day to night

depending on the time of day. Simply knowing when the sun will set can provide a good insight into the expected forecast for a given area. The setting sun reduces the differential heating, and this can spell an end to ongoing thunderstorms, particularly in tropical locations or near the sea.

The Moon

The other body in motion in our part of the solar system is the moon, which orbits the Earth every 27 days. While it doesn't impact the day-to-day weather very much (though during a full moon, temperatures at the poles have been documented to rise about half a degree), the moon does cause the daily ebbs and flows of ocean tides. It is important to be aware of the tides during heavy rain or tropical systems, as flooding can be exacerbated by the already-high waters.

The Atmosphere

The most liberal definition of the atmosphere includes roughly the first 6,200 miles above the Earth's surface. Nearly all meteorologists, myself included, would label the top 6,120 miles as outer space. Most satellites are well within Earth's atmosphere according to this definition.

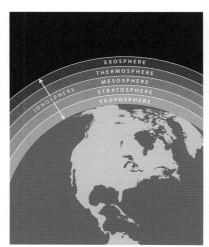

The layers of the atmosphere

There are five main levels to the atmosphere; from top to bottom, they are: the exosphere, the thermosphere, the mesosphere, the stratosphere, and the troposphere. The exosphere and the thermosphere are home to the Northern Lights but have next to no atmosphere and are not of much interest meteorologically. The next two layers down—the mesosphere and stratosphere—are also generally of interest only because of their temperature profile. Almost all of our weather occurs in the lowest layer of the atmosphere, the troposphere, which covers, on average, the bottom 7 miles of the atmosphere.

The Troposphere: Where the Weather Happens

The troposphere is where almost all of the weather we can see occurs, and where nearly the entire human race will spend their lives. (Even airplanes fly in the upper troposphere or the lower stratosphere).

The troposphere is marked by steadily decreasing temperatures as altitude rises. For example, at 15,000 feet, the air is colder than closer to the ground, perhaps in the single digits above zero. When one ascends farther, into the lower stratosphere, at the cruising altitude of most jetliners (around 35,000 feet), the temperature is often around -40 F and colder. There are two reasons for this. The first is that the sensation of temperature is based on molecules striking the skin or a sensor. Because the atmosphere is so thin at such altitude, there are fewer molecules that can deliver heat. The other is that the Earth's surface traps the sun's heat, and it radiates warmth into the atmosphere, heating the lowest levels more effectively than the upper levels.

At the very top of the troposphere, there is a boundary called the tropopause, where atmospheric temperatures begin to rise because there is less exposure to the turbulence of the lower levels. The chemical composition of the air in the tropopause (and the air in the stratosphere above it) is also different than the air below it and includes more ozone, which effectively traps heat. During the strongest thunderstorms, quickly rising thunderstorm clouds rise to the tropopause. In such cases, you can see exactly where the tropopause is because the cloud will flatten out at that level. If there is a particularly strong updraft,

A thunderstorm's overshooting top

the cloud will build above that flattened area, actually going into the stratosphere. This is called an overshooting top.

The stratosphere is a region of rising temperature, thanks in large part to the presence of the ozone layer, and it continues upward until the atmosphere is so thin that the observed temperature starts dropping again. This level is called the mesosphere, and beyond it are the mostly empty thermosphere and exosphere.

A vintage barometer

Pressure, Temperature, Humidity, Density

Barometric Pressure

Barometric pressure (also known as atmospheric pressure) is essentially the measure of the weight of the air above a location. It can be measured in bars, as is done across most of the world, or in inches of mercury, which is most common in the United States. As a measurement, referring to inches of mercury is not that helpful for weather analysis, but it's an easier way to explain the function of the barometer. Inches of mercury is the amount of mercury displaced by the weight of the air on top of it. With all that said, millibars are the unit most commonly used by meteorologists the world over. I'll be sure to reference both when necessary. Standard atmospheric pressure is 29.92inHG (1013mb).

The formula for calculating pressure is a bit too complex to mention here, but generally speaking, there are three main variables that can change in an air mass—pressure, density, and temperature. All three of those variables are interrelated, and together they help produce the weather we see each day.

Temperature and Density

Temperature is a measure of how energetic given molecules are. Warm temperatures are the result of more-active molecules, while cold air means that molecules are less energetic. Warm, active molecules need more space; this is why steam expands in a teakettle, for example. Cold air, on the other hand, is denser. This is why a blown-up balloon brought outside into the cold will look like it is deflated, even though the same amount of gas is present. The temperature changes simply made the gas inside denser, and it needs less space to move around

How Temperature Affects Pressure

To understand pressure, it's helpful to keep in mind two simple facts: Warm air has a lower density, and cold air has a higher density. Because barometric pressure is the weight of the atmosphere on top of a particular location, cold, dense air in a column above a particular spot means high pressure at the surface; relatively warm, less dense air in that same column would mean low pressure.

We often see pressure fall ahead of storms because the area of a cyclone (more on these on page 25) that is the most favorable for storm development is also the slice of the system that is the warmest (called the warm sector). It lies south of a warm front, but to the east of a cold front. The passage of a cold front means the introduction of cold air, and the pressure begins to rise.

Moisture, Humidity, and the Dew Point

From clouds, rain, and snow to humidity, the amount of water present in the air is an essential component of any forecast. The amount of moisture in the air is referred to as humidity. **Relative humidity** is the amount of moisture in the air relative to a particular air mass's capacity to hold water.

A good measure of how a day will feel is to look at the dew point. The dew point is the temperature at which moisture in the air can begin to condense into water droplets. If the temperature and dew point are close together, fog can be expected.

While relative humidity can tell us how much more moisture the air can hold, the dew point can actually indicate how the air will feel because it relies only on the amount of water in the air and is not calculated relative to the temperature. In terms of comfort, a dew point of about 60 degrees is a good benchmark for what feels humid, and a dew point of 70 degrees is downright oppressive. When reading or viewing a forecast or weather observation, look for the dew point, as it will be far more informative than the relative humidity.

Often, when people experience uncomfortably warm, sticky weather, they describe the weather as humid and say the air feels heavy. In fact, the opposite is true; water molecules are technically lighter (in terms of atomic weight) than the other constituents of air.

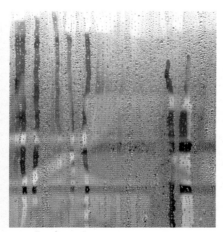

Condensation

Condensation

A crucial trait of water is its tendency to condense. A given air mass has a maximum amount of moisture that it can hold, and this amount depends on temperature. As air parcels rise, they cool, and eventually they become saturated —unable to hold any more water. When this happens, the excess moisture looks for anything to condense onto, such as particulates in the air or preexisting ice crystals. This happens at the ground level too. In fact, this is how we get dew—when the air cools down, the excess moisture is deposited on the ground, on your car, and so on.

Cloud Formation

All clouds are formed by rising air (also known as an updraft). Because there is more air at lower levels of the atmosphere, there is also more water vapor, and more particulate matter for water to condense upon. As air rises, that moisture and those particulates reach cooler air, the temperature meets the dew point, and moisture collects on the particulates and coalesces into large drops. A cloud is a collection of these large drops (or ice crystals, if the layer is cold enough). Rising air can originate at any level of the atmosphere, and the amount of moisture in the air directly impacts the height and type of cloud. For a field guide to clouds, see page 47.

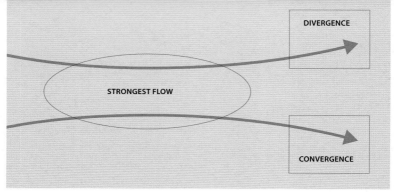
Divergence and convergence

The Bigger Picture

Atmospheric Current and the Jet Stream

In the middle latitudes (23–66 degrees north and 23–66 degrees south, approximately) the air generally flows from west to east, which is also the direction that Earth rotates. The strongest flows follow the strongest temperature gradients. A gradient is a change in a particular variable over a geographic area; a temperature gradient represents the change in temperatures over a certain distance. Areas with a strong flow are better known as jet streaks and the overall course these streaks take through the atmosphere is known as the jet stream.

Variation of the Jet Stream

The jet stream and jet streaks are generally found between an altitude of 30,000 and 40,000 feet, but this can vary with temperatures at the lower part of the atmosphere. Jet streams nearer the poles can be as low as 20,000 feet, while the jet stream can be more than 50,000 feet up if its path takes it close to the equator.

How the Jet Stream (and Jet Streak) Works

In the Northern Hemisphere, if you were to stand facing in the direction that the jet streak was flowing and you were located at the end of the jet streak, where airflow was slowing down, you'd see that the air to the right of the jet streak converges (comes together); you'd also see that the airflow to the left diverges (moves farther apart), as it is seemingly pulled toward the pole by the Coriolis force (the apparent force placed on air masses and currents by the rotation of the Earth). Therefore, the air on the left exit of a jet streak has lower pressure, and air rises to fill in this gap, creating clouds, rain, snow, and thunderstorms. If air converges at the jet stream level, it sinks, preventing most clouds and precipitation from forming at the surface level.

Convergence and divergence are closely tied to rising and falling air masses on a large scale. Areas of jet stream divergence are correlated with areas of low pressure at the surface, just as convergence in the jet stream is related to areas of high pressure at the surface. At the surface and lower levels of the atmosphere, density and pressure are a function of temperature, but at the very top of the atmosphere, divergence means that air molecules are becoming more separated, and there is simply less air in a column and less overall weight to cause a rise in barometric pressure.

A Good Way to Get to Know the Weather

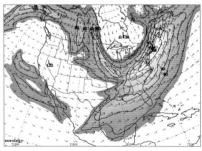

A jet stream map (note the U shape) during a polar vortex outbreak

Because of the outsize role it plays in generating the weather, becoming familiar with the jet stream forecast is a great way to get insight into the country's weather. The jet stream's strength is dictated by how much the air masses change, and this difference is a good indicator of the strength of weather features at the surface. If there are drastic differences between air masses at the surface, the jet can be more than 250 miles per hour, but to qualify as a jet streak, it need only have 50 mile-per-hour wind speeds.

Reading a Jet Stream Map

Meteorologists always take the jet stream forecast into account, as its shape reflects the air masses and where they come together. If the jet stream makes a U shape, the chance for a stronger storm system increases; this occurs because divergence is increased. Conversely, an inverted U shape would mean more-tranquil weather at the surface because of increased convergence at the jet stream layer.

In meteorological terms, that U shape is called a trough and the inverted U shape is referred to as a ridge. You might have heard of a dome of high pressure over an area, in reference to pleasant weather across a broad region. This means that there is ridge-like arc of the jet streak to the north of such a region that is deflecting away all the bad weather. Generally speaking, the bad weather at the surface lies below a jet. In a trough, it is more likely on the eastern side of the U-shape and can be just to the east of the trough itself. To the north (inside) of the trough, one can expect to find noticeably colder air. Conversely, to the south of a ridge, one can expect hotter, potentially humid air, with the threat for showers and storms found on the north side of the jet streak forming the ridge.

Seasonal Variations

In the winter, over the course of a few dozen miles, surface temperatures can range from the 50s to well below zero. In the summer, temperatures generally range from about the 90s to the 70s when a cold front passes through. Jet streaks over the United States, as a result, tend to be significantly weaker in the summer than in the winter. The stronger divergence associated with the winter jet stream allows for larger storms, both in terms of geography and intensity.

Low-Pressure Systems (Cyclones)

When the pressure is lower over an area, there is less air on top of a region. This means that there is warmer, less-dense air in the column of air above that location. Low-pressure systems are known as cyclones, systems where the flow rotates counterclockwise around the center of low pressure (in meteorology, this is known as cyclonic flow).

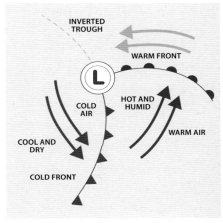

A low-pressure system

Generally speaking, meteorologists will only define a system as a cyclone if it is well organized, meaning only if there is a 360-degree circulation around a point, or if there are well-defined areas of cold or warm air advancing behind fronts. Warm fronts are typically found on the eastern side of an eastward moving cyclone, and cold fronts are found to the south; in between the two boundaries is the warm sector. The western flank of a cyclone features cold air.

A high-pressure system

High-Pressure Systems (Anticyclones)

High-pressure systems are also known as anticyclones. As their name suggests, anti-cyclones are essentially the opposite of cyclones; anticyclones rotate clockwise, rather than with the counterclockwise rotation of a cyclone. Such high-pressure systems have more air on top of a particular location, which means that the air is denser in the atmosphere above it. Cold air is denser, so while it may not be cold at the surface, there is cold air in most of the layer.

Areas with high pressure are often close to jet stream ridges because higher pressure leads to greater convergence at the jet stream level and therefore sinking air. Because of the sinking air, and because there typically isn't much change in temperature, it is more likely to be cloudless under high pressure, or at the very least, for clouds to be very thin.

Cyclones Often Bring Weather with Them

In a low-pressure system, there's divergence aloft, and air flows into the area above the low pressure point to fill in that gap. When that air is less dense and more buoyant, the air will tend to rise, leading to clouds and a greater potential for showers and thunderstorms. The flow around a low pressure system moves counterclockwise, most noticeably around the exterior of an area of low pressure, but the strongest winds tend to be in the center of the system. Hurricanes, the most notorious of cyclones, demonstrate this well, with counterclockwise windflow that's strongest near the eye, at the center of the system.

An Atlantic hurricane

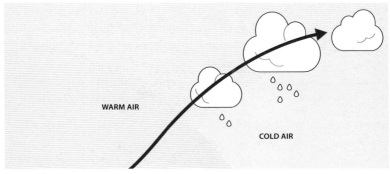

A warm front

Temperature Gradients and Fronts

When temperature gradients are sharp—when there is a great deal of temperature variation over a short distance at the surface—those gradients are called fronts. If the air mass is moving, we would call it either a warm front or a cold front, depending on the air temperature behind the front relative to the region it is moving into. Both types of fronts or boundaries have different characteristics and different effects on the people experiencing them.

Warm Fronts

Warm fronts represent the leading edge of a warm air mass that is moving into a colder region. Because warm air can hold more water than cold air, warm fronts typically also indicate incoming humidity, especially in the summer months. Warm fronts usually travel from the south or southwest, where it's generally warmer.

Warm air tends to be unable to push cold air out of an area. Remember, cold air is denser than warm air, which means it stays at the surface and is harder to push around. Because the cold air is entrenched, warm fronts only gradually change the temperature.

If the warm front produces any precipitation, it will generally be in the form of scattered, light-to-moderate rain or snow. The updrafts needed to produce clouds or precipitation can originate at any level, but with warm fronts, they start higher up and don't have as much moisture to work with at the start; they also don't travel upwards as quickly. For these reasons, cloud cover provided by warm fronts tends to be higher.

An unfortunate consequence of this pattern is that warm fronts, under the right conditions, can create freezing rain and sleet, two of the more underappreciated weather phenomena that can occur. This happens when precipitation originates within or above a warm front and falls as liquid droplets into the cold layer below, where they can partially refreeze into an ice pellet (sleet). Sometimes the droplets freeze upon contact with a solid surface, falling as freezing rain.

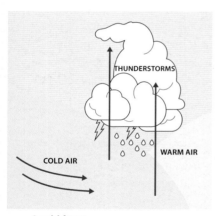

A cold front

Cold Fronts

Cold fronts represent the leading advance of a cold air mass into a warmer air mass. Cold fronts bring crisper air with reduced humidity, and conditions are often much breezier after they have passed, as the colder, denser air is unimpeded and more forceful as it trails the cold front.

Cold fronts tend to be more abrupt and noticeable than warm fronts, with a sharper temperature gradient across their boundary. This is, of course, because cold air is denser, and it doesn't filter in like warm fronts do; instead, cold fronts charge forward as they advance.

In meteorology, an air mass moving into a differing air mass is called advection. The momentum of a cold air mass advecting into a warm air mass causes the warm air to move, and it moves in the direction it has a tendency to move: up. This forced upward movement is unlike that seen in warm fronts. It is because of this that cold fronts tend to be associated with more-violent thunderstorm activity in the summer, though thunderstorms are possible with heavy snowstorms as well.

Because of how quickly cold air advances, any transitions between precipitation types occur quickly. Rain may be present before a cold front passes and snow may be seen after. Warm air at the surface and aloft exists so briefly during the passage of a cold front that there is little chance for melting and refreezing, so sleet or freezing rain in a cold front are usually inconsequential.

One nice thing about a cold front is how thoroughly it purges an atmosphere of any moisture. Because the cold air lifts the warmer, moister air parcels, the area a front passes over is left with cold, dry air in its wake. Not only does this lead to less humidity after a rainstorm in the summer, but it almost always leads to sunny skies for a period of time after the front passes.

How Big Is a Front?

You might be surprised to find out that there isn't really a correct answer to this question. A front itself is defined as the boundary between air masses. A cold, dry air mass tails a cold front, while typically, a moister, warmer air mass precedes it. The front is usually analyzed at the leading edge of the encroaching air mass, and it has an infinitesimal width. The length, on the other hand, is another story.

Understanding that there are other viewpoints, I view fronts as products of cyclones, with only one warm front and one cold front appended to a center of low pressure. The length of the front is proportionate to the strength of the storm system that is inducing the advection, or advancing cold or warm air.

Other meteorologists define fronts as any type of boundary in which a surface variable (temperature or dew point, usually) sees an abrupt change, which can lead to longer fronts being analyzed, or fronts that are not associated with storm systems or cyclones. This isn't an incorrect opinion, just a different one. I don't personally subscribe to it because I believe it is less informative and can lead to messier weather maps.

Occluded Fronts

Sometimes fronts overtake each other. When this occurs, it's known as an occluded front. (An occlusion is essentially where two things come together and result in sealing something off.) An occluded front means that warm air can no longer reach the center of low pressure. There are

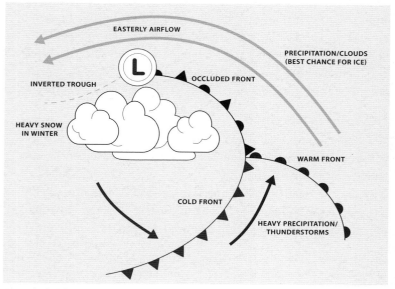

An occluded front

two types of occlusions: a cold occlusion and a warm occlusion. A cold occlusion means that after a front passes through, temperatures drop behind the front. A warm occlusion indicates that temperatures rise after the front passes through. Said a different way, a cold occlusion occurs when the cold air behind a cold front is colder than the air ahead of the overtaken warm front. A warm occlusion happens when the air ahead of the occlusion is colder than the air behind the original cold front.

Cold Occlusions

When an occlusion passes through, the responses are similar but more muted than if it were simply a cold or warm front passing through. Cold occlusions produce lighter, more general precipitation than true cold fronts, as air doesn't rise as swiftly until it reaches a higher altitude. The lingering warm air aloft tends to broaden the impact of the front, however, and the clean cutoff that is normally seen behind a cold front is not seen with a cold occlusion. Also, there is often residual cloud cover or even showers after a cold occlusion passes.

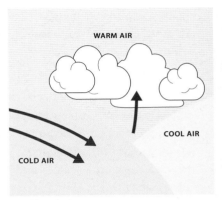

A cold occlusion

Warm Occlusions

Warm occlusions are actually a bit more favorable in some regards, as their cooler air is less apt to allow updrafts, and precipitation is reduced compared to a regular

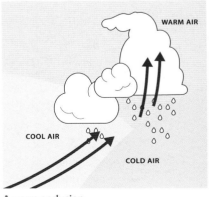

A warm occlusion

warm front. They occur when cool air from the overtaking front will override the cold air that's in front of the warm front, and subsequently force warm air upwards, and thunderstorms are more likely with warm occlusions than with warm fronts. So warm occlusions may feature less geographic coverage in rain, but that activity can be a bit more vigorous.

Dry Lines

Often compared to fronts because they are a barrier between air masses, dry lines are not associated with areas of low pressure. Most often occurring in west Texas, they result from the conflict between warm, moist air and the dry air of the desert Southwest.

Dry lines develop thanks to a westerly wind off the Rockies; the mountains block the wind, creating an area of low pressure on the opposite side. This is known as a lee trough, and it is an impetus for the warm moist air to attempt to move westward. Dryer air, remember, is going to be heavier than moist air because it is not filled with those light water vapor molecules, and as such, the moist air rises. Dry lines are a regular source of thunderstorms in west Texas, especially as night falls and the upper atmosphere cools, allowing for the moist air to become even more buoyant.

Dry lines don't advance like fronts; instead, they dissipate as airflow changes directions. They can produce some very strong thunderstorms, thanks to the available energy coming from the Gulf of Mexico, as well as the heat of a hot Texas afternoon.

Low-pressure Troughs

Not all areas of low pressure can be simply described as featuring cold or warm fronts. There are often troughs of low pressure embedded within a low-pressure system that emanate from the center of that system. Much like frontal boundaries, they feature converging winds at the surface; however, they are not tied to changes in air mass.

These troughs lead to precipitation with some regularity. On the surface, they can lead to updrafts, as the colliding air parcels have nowhere to go but up. In the area between the cold and warm front (the warm sector, as it's known) these troughs are called pre-frontal troughs and can often blunt the impact of cold fronts and send

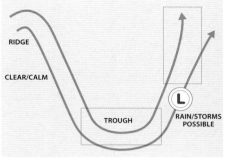

A low-pressure trough

hot, moisture-rich air up to form thunderstorms, leaving the front with less moisture to turn into showers or thunderstorms.

Inverted Troughs

When troughs are found north of the center of circulation, they are called inverted because they angle in the opposite direction most fronts and troughs find themselves in. Especially in the winter, east of the Rockies, inverted troughs can be prolific snowmakers, as they sit in the coldest part of the system and are able to draw moisture from bodies of water to the east, such as the Great Lakes or the Atlantic Ocean.

Free Convection

Sometimes, weather isn't the result of a larger mechanism, such as a weather front or a trough. A notable example of this is free convection. This occurs when air rises, thanks to its inherent buoyancy, often as a result of heat and humidity increasing into the afternoon. Atmospheric convection is the same phenomenon you see when a pot of water is boiling; the rising heat causes bubbles to rise upward from the surface and into the atmosphere. When free convection occurs, it can give rise to anything from turbulence for nearby aircraft to storms, including powerful thunderstorms.

Updrafts

A crucial component of convection, be it free convection or convection induced by a cyclone, is rising air, called an updraft. Depending on the magnitude of the rising air (determined by how warm the air is compared to the air above it), updrafts can lead to more low-level moisture merging with moisture aloft, leading to fatter rain drops or growing hailstones. Rain or hail can only fall when it is heavy enough to overcome the strength of the updraft.

An updraft

A looming storm

North American Weather Systems of Note

Once you have a grasp of the basic concepts of meteorology, it's not hard to put them into practice. In North America, over the course of a full year, a number of familiar storms and patterns emerge. From tropical storms and hurricanes to Nor'easters and Alberta Clippers, you've likely already heard of some of them. Here is a brief rundown of a few of them and how they work.

Nor'easters

One type of storm that makes a lot of headlines when it develops is the Nor'easter. One might believe that the name comes from the location it typically impacts, the northeastern United States, but it actually refers to the northeasterly winds that batter the coast as the center of the cyclone moves by.

The Nor'easter begins its rapid development in the southeastern United States, or more accurately, just off the coast. Nor'easters can often be some of the strongest systems found outside of the tropics, thanks to the two aids to their development, the Appalachian Mountains and the Gulf Stream, an oceanic current off the East Coast, which is a channel of warm water that helps to guide weather features that can feed off of the uncharacteristically warm water.

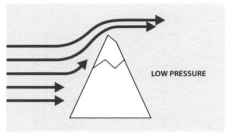

A lee trough

The Appalachian Mountains help produce such storms because mountain ranges often produce what is known as a lee trough. The basic concept isn't hard to understand. Because barometric pressure is essentially the weight of all the air on top of a certain point, if air is attempting to move through mountains, the air column is stretched out and some of it will be unable to make it through the range. If there is a west wind through a mountain range, then east of the mountains, there will be an area of low pressure essentially generated by the presence of a mountain range. These low-pressure areas can often give rise to storms.

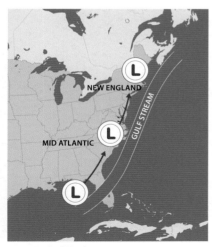

Typical Nor'easter movement

Many of the characteristics of a Nor'easter are similar to those of a hurricane, but unlike hurricanes, they also have the advantage of the jet stream's support. Since the jet stream plays a role in the development and sustenance of East Coast weather systems, there is cold air available for a Nor'easter to make use of, and this is a fundamental difference between a Nor'easter and a hurricane. It's also what gives Nor'easters their reputation for producing heavy snow on the cold, northwestern side of these storms, usually along the Eastern Seaboard. Because Nor'easters form over the ocean, they are also packed with moisture and can produce prolific snowfall totals, in conjunction with strong winds.

Because of their tight circulation, and the fact that they are effectively fueled by warm water, satellite images of Nor'easters often look quite similar to hurricanes. Often, they even develop traits usually found in hurricanes, such as an eye and numerous bands of precipitation.

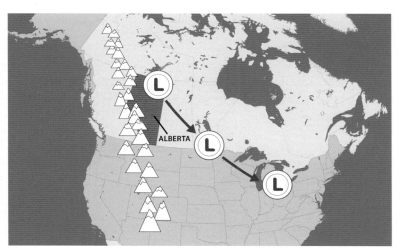

Typical Albert Clipper progression

Alberta Clippers

The Rocky Mountains certainly play a role in the development of storms in North America. There are two types of storms that you have perhaps heard of that are generated with the aid of the Rocky Mountains: Alberta Clippers and Panhandle Hooks. The **Alberta Clipper** is more common, and it develops in the lee (the side sheltered from the wind) of the high peaks of the Canadian Rockies. Alberta Clippers develop in areas where the jet stream only produces a slight trough (also known as a low-amplitude trough). The amplitude of a jet trough is directly correlated to the strength and organization of an area of low pressure, and as a result, Clippers don't tend to have very low central pressure, nor are they terribly well organized, sometimes lacking well-defined warm or cold fronts.

Clippers also don't have access to as much moisture as a system that organizes farther to the south, and they often only provide a quick batch of snow to the Canadian Prairies, Upper Midwest, and western Great Lakes; they only really become major snowmakers as they start pulling moisture from the Great Lakes and the Atlantic Ocean.

Another characteristic of Alberta Clippers is how swiftly they move. Because they have access to portions of the jet structure with more energy, they charge through the Plains more quickly than more-typical systems. Since they tend not to linger for very long, the threat of a bunch of snow is further reduced.

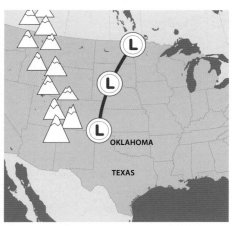

Panhandle Hooks

Panhandle Hooks (aka Texas Hookers)

While low pressure can develop anywhere just to the east of the American Rockies (particularly in Colorado), another significant area where storms develop is in the Panhandle areas of Texas and Oklahoma. Less frequent than an Alberta Clipper, this type of storm system is known as a Panhandle Hook or a Texas Hooker.

Panhandle Hooks are more unusual because they require the support of a much higher amplitude jet trough, which pulls in cold air all the way from Canada to west Texas. What Panhandle Hooks lack in frequency, they make up for in intensity.

Given the sharper jet trough required to create the Panhandle Hook, these storms are usually better organized, with more-easily discerned cold and warm fronts, gustier winds, and a lower central pressure. Their more-southerly origin also lends them access to the warm waters of the Gulf of Mexico, and, as the system tracks northeast, Panhandle Hooks can be the most prolific snowmakers of any given winter season for the Plains.

Panhandle Hooks tend to track along the southwest-to-northeast side of jet troughs, which brings them northeast towards the Great Lakes over their life span, tracking directly through the central Plains and the Mississippi River Valley. As they become better organized, these can be vast, and they can impact the country from border to border.

Polar Vortex

The polar vortex is a real meteorological phenomenon that has recently caught the attention of the national media. In the process, it was over-hyped to the point that it seemed simply like hyperbolic nonsense. In fact, the polar vortex is a real, ongoing phenomenon, no matter when you are reading this.

A vortex is simply fluid moving in a circular motion. The polar vortex, then, refers to the polar jet stream spinning around the North or South Pole. That's it! Of course, the polar jet stream holds back the coldest air the planet has to offer; the polar vortex earns its notoriety from recent instances when the polar jet sank far enough south to impact highly populated areas.

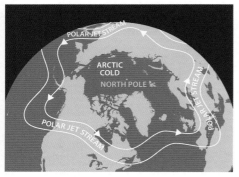

A polar vortex outbreak

The most incredible thing about the polar vortex is not that it exists, but how routine its appearances in the continental U.S. actually are. When the polar vortex lingered for weeks in the United States in 2013–14, it gained extra notoriety, but it usually makes a visit to the U.S., particularly in the Plains, at least once a winter.

Bomb Cyclone

One of the more-recent media buzzwords is bomb cyclone, and viewers may conclude that it is a term invented to get web clicks or TV viewers. In fact, like the polar vortex, it is an actual phenomenon with a specific definition within the meteorological community.

A bomb cyclone is an area of low pressure that drops by at least 24 millibars of pressure over 24 hours. The term has two explanations. First,

Lake Michigan during a bomb cyclone

the pressure drops like a bomb. The second ties the phenomenon to its technical name: explosive cyclogenesis, a cyclone that develops extremely quickly.

The kinetic energy involved in such a developing system means that winds flowing into the feature will be incredible, leading to a violent clash of air masses. These storms are most typically found in the transitional seasons, particularly from late winter through spring. Bomb cyclones often lead to prolific snowfall or severe weather, and often both!

This term came to national attention recently after a series of strong storms battered the East Coast. Explosive cyclogenesis is more likely over the sea, as landmasses produce friction that slows the development of cyclones, so it isn't unusual to find bomb cyclones in coastal areas, though they are infrequent. They can develop over the Plains as well, with the help of the Rocky Mountains, and can produce the strongest severe weather outbreaks of the season.

Blizzard warnings during a recent bomb cyclone

How to Read a Weather Map

The black lines on a weather map represent "isobars." The prefix "iso-" means "equal" and bars or millibars are the measure of atmospheric pressure. An isobar is therefore a line of equal pressure. Lower numbers represent lower pressure and higher numbers reflect higher pressure.

The blue line with triangles on it is a cold front, with the triangles pointing in the direction in which the front is moving. The red line with half circles is a warm front, with those half circles pointing in the direction in which the front is moving. Note that the fronts lie in areas where the otherwise circular rings around the center of low pressure are contorted. Other distensions in these concentric circles represent troughs and are likely areas of precipitation.

Though not used in this example, other common features on surface maps are stationary fronts, represented by an alternating line of blue triangles and red half circles pointed in opposite directions; occluded fronts, which are purple lines with half circles and triangles all pointing in the same direction; and brown or purple dashed lines, which are catchall symbols, either for surface troughs or dry lines.

WEATHER PHENOMENA YOU CAN SEE

Now that we have a simple understanding of the basic physics of meteorology, let's start looking at the more practical side of the weather, and how can you prepare and adapt to the weather, based on what you are seeing outside, on television, or on social media.

We'll start with some of the things you may see when looking up at the sky and how you might be able to determine what's going on, simply with a glance at the clouds.

45

A perfect day

CLOUDS

Clouds are the product of rising air, which causes water droplets within the rising air parcel to collide, coalesce, and accrete into larger droplets, until they reach the point that their presence can reflect and refract sunlight, making them visible to the naked eye.

Though this is how all clouds are formed, no two clouds are the same. Each cloud type forms as a result of a number of variables; these range from altitude and temperature to the strength of the updrafts. In all, there are 10 primary cloud types; by being able to recognize these 10 varieties, along with a few other notable cloud types, you'll be able to start understanding the basics of the broader weather patterns in your area.

Of the types of clouds that you are most likely to see every day in the middle latitudes, there are three base types of clouds—cirrus, cumulus, and stratus—and two helpful modifiers: nimbus, meaning convective, and alto, meaning mid-level. Most of the 10 cloud types below feature some combination of those five basic roots.

Often, these root terms are used as blanket terms for all clouds of the subfeature, as they will be in other pages of this book. For example, "nimbus clouds are efficient rain producers" where "nimbus" can mean cumulonimbus or nimbostratus, or "warm fronts produce broad swaths of stratus" would refer to all clouds of the so-called "stratiform" family, like true stratus, nimbostratus, altostratus, and so on. As with the rest of meteorology, there are no sharp lines that separate these categories, and meteorologists often strive to be as broad as possible without being unhelpful.

Cirrus

Cirrostratus

Cirrus

What They Look Like: Cirrus clouds are wispy clouds found extremely high in the sky. In fact, cirrus clouds can often be found above commercial airliners at cruising altitude.

What They Say About the Weather: Because they are so high, they rarely have much to do with the weather at the surface, and they are generally indicative of fair weather. If you are seeing cirrus clouds, either the intervening layers are so dry that clouds can't form, or cold air is found fairly low in the atmosphere.

Cirrostratus

What They Look Like: Cirrostratus clouds are fields of clouds that are fairly wispy and at a very high altitude. Cirrostratus clouds don't really look like clouds at all; they look more like haze. They're easy to spot when they form a halo around the sun and, especially, the moon at night.

What They Say About the Weather: They form because of high moisture levels at the altitude (20,000 feet and higher) of cirrus clouds; this can indicate the advance of a warm front, as the cool, dense air remains at the surface with the more buoyant air hanging out aloft. Because the haziness usually precedes a warm front, the haze of cirrostratus can mean that clouds will eventually thicken, and stratus clouds at lower levels are soon to follow.

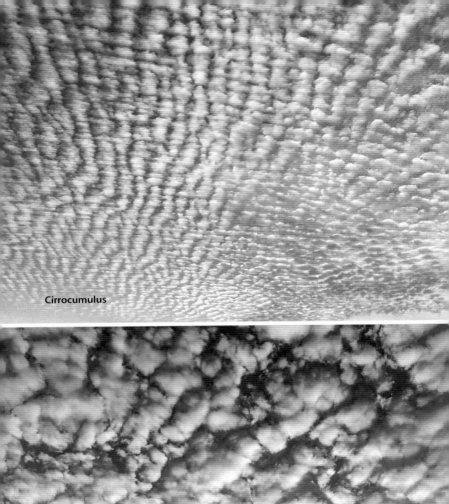

Cirrocumulus

Altocumulus

Cirrocumulus

What They Look Like: Cirrocumulus clouds, unlike stratocumulus clouds and altocumulus clouds, are so thin as to be almost grainy in appearance.

What They Say About the Weather: Sometimes cirrocumulus clouds form a striking herringbone pattern, often called a mackerel sky; when you spot a cirrocumulus mackerel sky, with high, wispy clouds, you're seeing clouds blown by strong winds at that cloud level; they point in the direction that the cloud decks are moving and can sometimes indicate that a front is on its way. Because of the height of cirrocumulus clouds, a broad area with this herringbone pattern may indicate the position of the jet stream.

Altocumulus

What They Look Like: Altocumulus clouds are sheet clouds, like cirrocumulus, and have smaller puffs of cloud than stratocumulus clouds, but larger ones than seen in cirrocumulus clouds. Because the clouds are lower, they may appear to have more definition than cirrocumulus.

What They Say About the Weather: Altocumulus clouds suggest shallow convection, which means that there are weak updrafts through a thin layer of moisture. In short, they mean that the weather is in pretty good shape for the time being. Sometimes, a sheet of altocumulus will have a herringbone, mackerel sky pattern, which indicates a brisk mid-level wind, often behind a recently passed front.

Altostratus

Nimbostratus

Altostratus

What They Look Like: Altostratus are thin clouds, though they cover the entire sky. They may appear bluish in nature, and this is because the sun and sky can filter through the thin cloud.

What They Say About the Weather: Altostratus are formed because there is a great deal of moisture at the middle levels. They indicate instability in the atmosphere and suggest that rain is on the way.

Nimbostratus

What They Look Like: The difference between nimbostratus and run-of-the-mill stratus is that the extra warmth of the season makes the rising parcels more unstable, creating some of the bubbling at the top of the clouds that you will also see in cumulo-nimbus clouds. Nonetheless, from ground level, nimbostratus are often dark gray and featureless and rarely photograph well.

What They Say About the Weather: Nimbostratus usually only exist around fronts, particularly summertime warm fronts but also cold fronts after the stronger, but more isolated, cumulonimbus have matured into a broad field of showers and thunderstorms. Nimbostratus clouds don't necessarily predict anything. Instead, they tell you something you've probably already figured out: It's raining, and it's probably been raining for a while.

MID-LEVEL CLOUDS (ABOUT 6,500–20,000 FEET)

Stratocumulus

Stratus

Stratocumulus

What They Look Like: Like altocumulus and cirrocumulus clouds, stratocumulus clouds consist of sheets of puffy clouds at the same altitude. Stratocumulus tend to have larger puffs of cloud.

What They Say About the Weather: They indicate that shallow convection (warm air rising into cooler air) is occurring.

Because they represent shallow convection, they indicate that there is a measure of stability above and below the field of clouds. When they appear by themselves, they are good indicators of pleasant weather. When these sheets of cumulus clouds appear with other clouds, the other clouds should be used as the first indicator of what kind of weather to expect.

Stratus

What They Look Like: Stratus clouds are the thick, even-looking clouds that hang low in the atmosphere.

What They Say About the Weather: Unlike cirrus clouds, stratus clouds coalesce after only a short rise, due to an abundance of moisture near the surface. Stratus clouds are most often formed in systems where updrafts aren't as strong, and often the stratus layer isn't terribly thick. The widespread rising motion of these systems, however, ensures that the stratus covers a broad territory.

Stratus are most likely to produce light drizzle or snow showers. Because they are found where air parcels have more trouble rising, there is less opportunity for embedded water droplets to grow into large enough drops to fall through the rising air.

Cumulus

Cumulus

What They Look Like: Cumulus clouds can certainly develop at low levels, which is why they are included here, but they are also possible in the middle levels. They can have bases as low as 1,000 feet or as high as 15,000 feet.

Cumulus clouds are the clouds you drew as a child: puffy clouds floating across the sky. They are thick enough to blot out the sun when they pass by and will cast shadows on the ground. These clouds often develop when the weather is warm; when this happens, updrafts can be a little more vigorous, and additional moisture can make them denser and more cotton candy-like in appearance.

What They Say About the Weather: When found in isolation, and when it's warm, they are fair-weather clouds. In isolation, the greatest threat from cumulus clouds is a picturesque back-drop to a pleasant afternoon. As cumulus clouds grow larger and gain company, they become rain clouds. With advancing cold air, more concentrated rising air leads to more collisions and coalescence, leading in turn to more accretion and bigger clouds with bigger drops. Therefore, when the sky is filling with cumulus clouds, it's a sign that a change in the weather is coming, at the very least, but it may also be a sign of impending rain showers.

Cumulonimbus

Cumulonimbus Clouds
(and How Thunderstorms Form)

What They Look Like: As noted earlier, the word "nimbus" is a modifier of other cloud names; the strict definition of a nimbus cloud is a large gray rain cloud. Nimbus clouds can produce either rain or snow; nimbus clouds tend to darken the sky more than their counterparts, and they get their darker gray appearance because of their increased moisture content. This can be caused by two different factors; either the available moisture at the surface and within the rising air parcel is abundant, or the air being forced upward into it (whether from an advancing cold front or intense surface heating) is strong enough to cause more-frequent collisions and coalescence.

Cumulonimbus clouds are likely the most famous types of nimbus clouds. A cumulonimbus cloud has a broad base because air is being drawn toward the primary updraft before being forced upwards. The strong updraft creates localized low pressure underneath the storm, which causes the storm to act like a vacuum, sucking up the warm, moist air around it, with collision and coalescence happening at the base of the cloud over a broader area than the main updraft area. At the surface, this can also lead to brisk straight-line winds, both preceding the thunderstorm and occurring again after it departs.

What They Say About the Weather: Often called towering cumulonimbus for the heights that they can reach, these clouds mean only one thing—thunderstorms. The violent updrafts that cause these storms tend to occur when the surface (or the area just above it) is warm and humid. These clouds often coincide with features that lead to a lot of rising air, such as a cold front,

A thunderstorm cloud

a dry line, or a surface trough, but cumulonimbus are possible with free convection as well.

Especially in the evening, you can see lightning flashing within thunderstorms, from the top to the bottom. Thunderstorms emerge from the strongest updrafts and can offer some of the most dangerous, but also the most visually captivating, of all weather scenes.

Cumulonimbus can reach as high as the tropopause, which causes the top of the clouds to flatten, as temperatures actually become warmer higher in the atmosphere, thanks largely to the stratosphere's ozone, and parcels are unable to rise any farther. The strongest updrafts can lead to a few bubbles over this flat portion, called an overshooting top. The flat top of a cumulonimbus and its broad base leads to their nickname "anvil clouds."

Thunderstorms produce many of their own types of clouds that are found only within thunderheads. The following is an introduction to several of them.

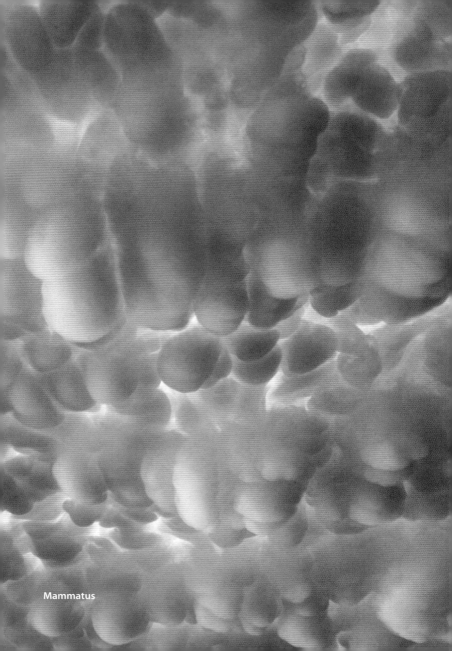

Mammatus

Mammatus

What They Look Like: Mammatus clouds resemble scalloped parcels of cloud hanging down from the base of a thunderhead.

What They Say About the Weather: While these clouds are found primarily in cumulonimbus clouds and are often associated with severe thunderstorms, seeing them is not necessarily a sign of bad things to come.

Mammatus clouds are something of an outlier in that they are associated with air moving *downward*, which means cold air emerging to the surface. (Most clouds form when air rises.) This occurs in two ways within thunderstorms. Some mammatus clouds form at the leading edge of a storm where warm air and cold air are mixing as the storm advances. The cold air pulled upwards falls out at the leading edge of a storm. Seeing mammatus clouds in such a case indicates that gusty downdrafts or even hail might be on the way.

The more common cause of mammatus is actually something less dramatic, and the thing that kills most thunderstorms: rain. Falling rain works to suppress updrafts, and as a result, it allows for cold air aloft to descend with the rainfall. This contributes to the gusty winds behind a thunderstorm, but it is also generally a sign that the worst has passed.

Shelf cloud

Shelf Clouds

What They Look Like: Shelf clouds are a low sheath of horizontal clouds that often outpace the broader thunderstorm cloud.

What They Say About the Weather: Unlike mammatus clouds, shelf clouds are almost exclusively seen before weather is about to turn severe. If you see one, get to a safe place. Shelf clouds are a sign that strong straight-line winds are on their way.

Shelf clouds form when downdrafts reach the surface and spread away from the initial downdraft. Found on the very leading edge of the thunderstorm, they are pushed forward by the storm's motion and momentum and by thunderstorm downdrafts, which are caused by falling rain redirecting and overwhelming updrafts.

Shelf clouds are often accompanied by fractus clouds, or scud—small, dark, fast-moving clouds that flow easily with the wind moving to or from the thunderstorm. The ominous appearance of the shelf cloud and the movement of the fractus clouds can give the appearance of a looming tornado. While a dangerous situation is present, fractus clouds and shelf clouds are not the clouds found when tornadoes are produced.

Wall Clouds

What They Look Like: Many people see shelf clouds and misidentify them as wall clouds. Shelf clouds are much longer and cover the full base of a thunderstorm, whereas wall clouds are a more localized lowering of the base of the cloud. Shelf clouds move parallel to the ground, in the direction of the storm motion, whereas wall clouds rotate around a vertical axis through the wall cloud. Scud (see page 65) is another type of cloud often confused for funnel clouds; scud clouds slowly rise to meet the base of the thunderstorm.

What They Say About the Weather: Wall clouds are perhaps the most notorious of all clouds found within thunderstorms. Wall clouds are the parent clouds to tornadoes, so this is one cloud you need to know how to identify. Funnel clouds, which become tornadoes, emerge from the base of wall clouds and descend down towards the ground; they **only** do this from wall clouds.

Wall clouds most often occur at the southwestern corner of a thunderstorm. This means that they are often rain wrapped and difficult to see when they are advancing toward you; when observed from west or south of the storm, they are much easier to see. Not coincidentally, this is also where you will find most smart storm chasers.

Keep in mind that while the funnel is the most visibly apparent part of a tornado, it is important to heed warnings and treat wall clouds as if there is a tornado ongoing. Tornado touchdowns can be invisible to the naked eye, as the whirling wind column can exist away from the visible cloud structure. Dust and airborne debris may be the first way, aside from radar way to determine if a tornado is already ongoing.

A heavy rain

PRECIPITATION

There are three primary types of precipitation—rain, snow, and other types of frozen precipitation. They are produced by a number of different cloud types, and they also produce different visible effects when they fall.

Rain

Rain is the most common form of precipitation. When air rises, it begins to cool because the temperature is lower at higher altitudes. As the air cools down, it loses its ability to retain moisture; the moisture then condenses out, latching onto any surface it can find—often a tiny particle of dust or debris, or an existing ice crystal. Because this process occurs high up where the temperatures are very low, many raindrops actually start out as snow.

Clouds remain suspended in the air because updrafts of warm air hold them up in the atmosphere. Because of these updrafts, the process of condensation can continue; this causes rain droplets or ice crystals to collide and coalesce. If there is enough available moisture, those droplets/crystals will get large enough that they begin to fall, despite the updrafts. As the precipitation falls, it enters warmer air; if the surface temperature is above freezing where it is falling, the precipitation will likely fall as rain. If it is below freezing, it'll fall as snow.

An updraft

Which Clouds Produce It: Rain always comes from cumulus or stratus clouds. As a general rule, because cumulus clouds are taller, they will produce heavier rain and larger drops. Stratus clouds are more apt to produce long-duration drizzle. Again, because they tend to be taller, cumulus clouds put more rain droplets between the sun and the person on the ground, so they can appear to be darker, while stratiform clouds are gray, allowing a little sun to filter through the drops.

To make a broad generalization, the heaviest rain comes from the darkest clouds. Clouds are darker because less light filters through the clouds, and unless it's dusk, the only thing that will filter out sunlight is additional moisture within clouds. It's not always a perfect metric, but it is a fair enough estimate of a cloud's rain-producing potential.

Other Precipitation Notes: Updrafts are strongest where heating at the surface is warmest. As a result, warm surface temperatures leading to strong updrafts will typically lead to heavy rain. In addition, because warm air can retain more moisture, in nearly every scenario, rainstorms produce more overall liquid than snow storms. (As a very rough general rule, 1 inch of rain is equivalent to 10 inches of snow, though the actual amounts vary because of a number of factors.)

Oddly, rainstorms can sometimes quash the very storm that created them. Storms with heavy, dense raindrops can actually suppress updrafts. Unless storms are on the move, heavy rain usually works against an active rainstorm or a thunderstorm.

Sometimes, updrafts can occur in drier environments where there is moisture aloft for clouds to develop and even for precipitation to begin to fall, but close to the surface, the atmosphere is very dry. In such cases, the falling raindrops evaporate into the surrounding air, and they never reach the ground. Usually when this happens, the rain is falling from a broken cloud layer, which means there is sun peeking through the

Virga

clouds, and you are able to see the rain shafts from a distance. If the rain shaft doesn't reach the ground, it's known as virga.

Ironically, the thing that allows us to see virga—the sun backlighting a cloud—can also confound it. If there is too much light behind clouds, one might see crepuscular rays. These parallel sunbeams shine through in between spots in the clouds, and they can often look like rain shafts or virga, especially when those rays are vertical. Of course, without beams of light from the sun, there wouldn't be enough light to even see the rain shafts themselves.

Hail

Hail

Hail is similar to other varieties of icy precipitation, but it is unique in that it is exclusively a warm-weather phenomenon. It forms in thunderstorms with strong updrafts when ice droplets fall and are forced back upwards. In this ping-pong-like process, they collect another layer of liquid, before falling back into the updraft. This continues until the hailstone is large enough that gravity can overcome the force of the updraft.

Large hailstones are a product of violent winds within thunderstorms. For a storm to stay active, it requires airflow into a storm; if the airflow isn't buoyant enough to form a rapid updraft, it begins to rotate. Though the presence of large hail—greater than about 2 inches in diameter—isn't a decisive indicator that a tornado will occur, it's a fairly reliable indicator that a tornado is possible.

The most common hailstones are about the size of a pea, with their rarity increasing with size. Storms that produce hail an inch or greater in diameter are considered severe because of the damage those stones can do to crops or automobiles. Hail that is any larger than an inch and a half can damage roofs and break glass. The worst storms can produce hail that approaches 4–5" in diameter, a size that is usually only seen once or twice a year, typically in the central or southern Plains. Hail of this size can be lethal for anyone caught outside, and it often causes problems for livestock too. If hail is being reported or expected, the safest thing to do is to get under something sturdy, preferably indoors, and stay away from glass or other breakable materials.

Which Clouds Produce It: Hail forms in the cumulonimbus clouds of thunderstorms.

Snow

Snow

If the area between the clouds and the surface is mostly below freezing, precipitation will fall as snow. Colder temperatures lead to crystallized flakes falling the whole way to the Earth, and colder snows, unless they are wind driven, often fall more gently to the surface. Because it's so light, however, this snow blows around easily and leads to drifting and visibility issues in open country in particular.

Water is a unique molecule. Molecules tend to get closer together as they cool, therefore taking up less space. Water, however, is different, and it begins to expand as it gets colder (near freezing). When it freezes, it crystallizes in the snowflake's familiar six-sided (hexagonal) shape.

Sometimes, snow falls when there's a layer of air that is above freezing near the surface. This produces slightly liquid snowflakes that are more capable of merging with other flakes, resulting in the fat snowflakes you might see in some storms. This merging process also disrupts the growth of individual flake patterns somewhat.

Regardless of the type of snowflake, individual snowstorms rarely contain as much moisture as even the most generic heavy rainstorm.

Which Clouds Produce It: Because snow falls when it is cold, it only falls from a few types of clouds. As there is little convection (warm air rising) present in the winter, snow rarely falls from cumulus clouds and is almost always the product of stratus clouds. The exception is within very strong storms, such as Nor'easters.

Partially because there is less vertical depth to snow clouds, and partially because of their lower moisture content, snow clouds are significantly lighter than rain clouds. Additionally, because air begins condensing more quickly when it's cold, snow-producing clouds tend to have lower bases than rain clouds. This is what makes snow days relatively gloomy: the low-hanging stratus clouds are a rather featureless matte gray.

Other Precipitation Notes: When areas of low pressure are very strong, the forward momentum of air forced around the center of the pressure system can induce a little extra convergence, producing a little bit more rising air. In these cases, there *can* be snow in cumulus clouds, and such storms almost always produce more-prolific snowfall totals. These are the same type of snow events that can produce the extra-rare thunder snow, which occurs when stronger than normal updrafts produce lightning within the colder environment. Snowstorms with thunder are always heavy snow producers.

When it comes to snow, and how it forms, mountain snows are something of an exception. To put it simply, air can't move through mountains, so it instead must go up and over. Technically speaking, this is known as mechanical or orographic lift. When this happens, the air rises and cools, and, as a result, moisture is wrung out of the air with incredible efficiency; most major mountain chains have a dry region on the downwind side of the range. These mountain-induced snowstorms, particularly in places like the Sierra Nevada Mountains near the California-Nevada border, can produce storms that provide the entire region with water for the next spring.

Freezing Rain, Sleet, and Graupel

Freezing rain, sleet, and graupel all form when warm air overrides areas with cold air; this is most common with warm fronts.

Which Clouds Produce It: Freezing precipitation almost always comes from stratus clouds that cover the lower-to-middle altitudes. Such clouds usually have a uniform consistency at their base. In very rare instances, such as with stronger systems, nimbostratus clouds can produce freezing precipitation, though in such events the responsible warm front is moving too fast for the icing to continue for very long in one place.

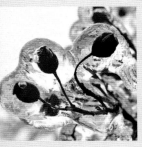

Freezing rain

Freezing Rain

Freezing rain is precipitation that falls through the atmosphere as liquid rain until it reaches the surface. As it falls, it encounters a thin layer of cold air above the surface, which cools it below the freezing point. When they form, ice crystals need something to freeze around. Aloft, it's usually a dust particle or another ice crystal, though when temperatures drop below 32 degrees only shortly above the surface, there isn't anything for the falling water droplet to freeze onto. Because a layer with particulates isn't available in this scenario, the water becomes supercooled and remains a liquid until it hits the ground; once it does so, it freezes on contact.

Identifying Freezing Rain: It's tough to identify freezing rain until it's too late. It will form a coat of ice over most surfaces, and it makes even walking down the sidewalk treacherous. Freezing rain can become destructive after only 0.10" of accretion or collection on a surface, as it weighs down branches, power lines, or other structurally deficient items. Even just a quarter inch of freezing rain over a region can paralyze whole cities.

Sleet

Sleet

Sleet occurs when water droplets fall through a slightly thicker cold layer at the surface than freezing rain. This causes them to freeze just before they reach the surface.

Identifying Sleet: The best way to identify sleet isn't necessarily even by seeing it. It's by hearing it. You know sleet is falling when you hear it bounce as it hits the ground; this happens because it has a hard shell, even if some of the core remains liquid. If sleet is large enough (maybe ⅛ inch in diameter), you will be able to see it bouncing off of hard surfaces. Sleet can fall on warmer surfaces and begin to melt and then refreeze, creating slick roads and sidewalks. Fortunately, it tends to fail to accrete on elevated surfaces, like branches or car doors.

Graupel

Graupel

Graupel is something entirely different. It starts as snow and falls through a layer with supercooled droplets, which accrete, or collect, on the snow and freeze. Some people refer to graupel as soft hail, because it isn't quite a solid ice chunk, but it doesn't consist entirely of frozen liquid. Of the three types of precipitation listed here, graupel is the largest.

Identifying Graupel: Graupel might initially be identified as hail, but it certainly won't be as loud as a solid chunk of hail bouncing off your house. You may also be able to crush it between you fingers when you handle it. Since graupel is a result of congealed snow and ice, it tends to pose dangers you would normally associate with snow, rather than with other types of freezing precipitation.

Dew and Frost

Frost

Because they don't form in clouds and fall to Earth, dew and frost aren't technically forms of precipitation (water vapor that condenses and falls to the ground). Overall, however, the condensation process that produces dew and frost at the surface is relatively similar. Just as rain and snow form when moisture condenses out as air cools down aloft, dew and frost form at the surface level when temperatures cool and the air loses its ability to retain as much moisture. The saturated air deposits excess moisture on grass, cars, and any other object it can find. When it's colder, frost forms.

The main difference is that, when condensation occurs aloft, the resulting water/ice droplets are scattered and diffuse, as there are not as many surfaces to attach to. However, there are many potential surfaces to attach to and cover; this is why dew and frost often cover vast areas, such as entire lawns or your car's windshield.

Dew is formed because of cooling, as is one type of frost called radiation frost, or hoarfrost. It is called this because overnight heat (usually) radiates out into the lower atmosphere and away from the surface. Advection frost occurs when the temperature drops quickly, as with a cold front, and can lead to icy spikes forming. Icing on aircraft is another form of frost, called riming, that acts like freezing rain, depositing in clear sheets on a plane's wing and the aircraft's nose.

Fog

Fog is a type of low stratus cloud. So low, of course, that it reaches the ground and obscures vision at the surface. There are different types of fog, caused by different means, which can give us a good guess as to where fog is often found.

Radiation fog

Radiation Fog

Keep in mind that clouds are caused when moist air rises into cooler layers, which leads to condensation and the clouds we see above. Radiation fog is a similar process, but in reverse. At night, heat from near the surface radiates into the higher levels of the atmosphere, causing surface temperatures to cool. The cooler surface temperatures lead to the condensation we would typically see aloft.

Advection fog

Advection Fog

Advection fog is the type of fog we see along the coasts, often as part of the marine layer, a daily sheath of fog and low clouds often seen on the California Coast west of the Coastal Range. It is caused when warm air moves over colder surfaces, like tropical air moving over the chilly Pacific, or warmer air moving over snow in the springtime.

Evaporation Fog

Evaporation fog is the result of cooler air sitting above a warm body of water. The bit of evaporation from the water immediately saturates the air above it, leading to the wispy fog often seen in the early hours above lakes and ponds.

Evaporation fog

Topographical Fog

Often, the geography of an area plays a role in fog development. Flow that runs into mountains and moves up the slope (a process called upsloping) will create fog or clouds on the mountainside. In valleys, cold air can stagnate, unable to move over or around the nearby hills. The cooler air, in concert with any moisture, can lead to stubborn valley fog.

Topographical fog

A strong tornado

SEVERE WEATHER PHENOMENA

Some weather phenomena only occur during storms or severe weather events. Lightning and tornadoes are obvious examples, but flash floods, blizzards, and heat waves can also pose dangers to both people and property. Here's how to recognize such conditions and stay safe in them. Of course, the best preparation is staying informed; for information on weather watches, warnings, what they mean, and how to find out about severe weather in your area, see page 108.

Lightning

Lightning

You couldn't have thunderstorms without lightning, which is a byproduct of the aggressive updrafts within nimbus clouds. Thunderclouds form as warm air rises very quickly (upwards of 60 miles per hour); as the air rises, warm air condenses out, and the resulting water droplets brush past each other, producing static electricity. Electricity, much like air pressure, seeks an equilibrium or balance, so all that electricity eventually needs to find an outlet, either as a lightning bolt that connects with another cloud, or as a bolt that strikes the ground.

Lightning takes the path of least resistance to connect positive and negative poles, which usually means that lightning stays within clouds, but when bolts reach down to the surface, they often like to find the most exposed, highest-reaching object on the ground, either the tallest tree in a forest, the tallest point of a building, or a lonesome boat in the middle of a lake. Often, of course, doesn't mean always. While the storms may be vast, lightning bolts are quite narrow, which means that the course a bolt takes to reach the surface might not include the absolute highest point of a region. Lightning also seeks the most conductive material, and it will bypass one tree for a more conductive type of wood or find a building with a metal antenna instead of one without. So if you hear thunder or see lightning, be sure to get someplace safe, preferably indoors. Just because lightning strikes a tree doesn't mean it can't finish its route to the surface by passing through someone standing near it.

On occasion, you may hear of heat lightning; this isn't a real meteorological term. It usually refers to lightning that is seen from a distance, without thunder. In such cases, what you're seeing is a storm that's producing lightning; it's just too far away for you to hear the thunder.

A waiting storm

Thunder

Thunder is the direct result of lightning; when a lightning bolt discharges, it is incredibly hot—as much as 50,000 degrees. This causes the air surrounding the lightning bolt to become superheated, and this air expands away from the bolt, creating the massive crack of a lightning bolt and the long rumbles of thunder that echo across the terrain.

Staying Safe

Thunderstorms are an odd weather phenomenon in that they have a built-in warning system—bolts of lightning that are visible for miles, and thunder, which is usually hard to miss. If Mother Nature gives you that warning, it is best to heed it and get indoors.

If you are unable to get to shelter, it is important not to be the tallest object in your area (e.g. standing in the middle of a fairway when golfing, or staying on a lake when fishing). Also, don't stand near or under other conductors because lightning bolts are capable of arcing from that object to you on their way to the ground. Trees are good conductors, and they therefore provide poor shelter from lightning.

If you are unable to get to safety and are in an open area, those that have been struck or nearly struck say that you can feel a rise in static electricity shortly before lightning strikes. If you have any reason to believe that a lightning strike is possible, the safest thing to do in an open field is to crouch and give yourself a low profile, while at the same time giving yourself less contact with the ground, which may return a charge back up to you if lightning were to strike nearby.

Tornadoes

A twister

Tornadoes are often mistaken for clouds, but technically speaking, they aren't a cloud type. Instead, they are a tight circulation of wind that emanates from the base of wall clouds. Often, what appears to be a cloud toward the lower portion of a tornado is actually dust and debris drawn into its path by its high winds.

Tornadoes occur in thunderstorms with very strong updrafts. Updrafts, despite their name, draw air from all around a thunderstorm, rather than just below. Updrafts at the leading edge of a storm tend to be stronger and, therefore, can cause the entire storm to rotate. This rotating column of air will start to reach down on the rear side of the storm, in a feature called the rear flank downdraft, a process that creates a wall cloud. When that rotating column reaches the ground, a tornado is born.

It's difficult to predict the direction a tornado will move, especially in the short term, but they will generally move in the same direction of the thunderstorm. They vary in intensity though their life cycle on the ground and are measured using the Enhanced Fujita Scale, ranging from EF0 (wind speed below 85 mph) to EF5 (winds of greater than 200 mph). Unfortunately, we can only estimate the strength of a tornado after the fact, using damage surveys.

TORNADO RATING
Enhanced Fujita Scale

RATING	WIND SPEED (miles per hour)	DAMAGE
EF0	65–85	minor roof, branches
EF1	86–110	broken windows
EF2	111–135	roofs off, large trees
EF3	136–165	homes damaged
EF4	166–200	homes leveled
EF5	200+	incredible damage

The enhanced Fujita scale

Staying Safe

If a tornado is believed to be in the area, go to an interior, windowless room on the lowest level of the house, and put as many walls between yourself and the outdoors as you can. With strong winds and tornadoes, it is possible to get stuck out on the road or in an open area unable to reach shelter, and the best thing to do is keep a low profile. Tall vehicles, like trucks or vans, can be blown over, and remaining upright exposes you to more debris. Stay low and out of the wind if you are unable to find shelter.

Tornado Myths

Tornadoes get more media attention in the United States than perhaps any other weather phenomenon. The dangers tornadoes pose certainly warrant such coverage, but tornadoes are also the subject of a great deal of misinformation, and that includes many safety tips that were recommended once upon a time but are now out of date.

Here are a few of the most common tornado safety myths:

- Open all the windows to equalize the pressure.

- Go to a southwest-facing corner of the house.

- If you are on a highway, take shelter tucked under an overpass.

As you may have guessed, none of these tips are good advice.

Why Opening Windows Won't Help: The idea of opening all the windows in the house developed because of the belief that tornadoes caused such localized low pressure that structures literally exploded outward as a twister passed overhead. The truth is that the pressure gradient found near a twister isn't nearly strong enough to do this. Instead, the damage is caused simply by the strong wind, which can be as strong as 250 mph, or by debris launched by those strong winds. The fact is, even if the pressure discrepancy was strong enough to eventually cause structural damage, it would exist only at the center of the tornado, so the winds at the outer edges of such a storm would obliterate any structure long before the interior/exterior pressure gradient. So instead of opening windows before a twister hits, close them, and then follow the tornado safety tips on page 84.

The "Southwest Corner" Rule Isn't Real: The logic behind the old belief that it's wise to move to the southwest corner of a structure when a tornado arrives is even more curious. The origin of this bad advice stems from the idea that all severe weather arrives from the southwest, but that's not true. Storms generally come from the west, but they are just as likely to come from the north or northwest. Of course, even if all tornadoes approached from the southwest, going to the southwest corner of the house wouldn't help. Tornadoes, of course, have rotating winds, and debris can come from any direction.

Do *Not* Take Cover in a Highway Underpass: If you are stuck outdoors when a tornado strikes, get to the lowest level you can (that's where the winds will be weakest), but do not take cover under a highway overpass. While there are several infamous videos of motorists tucking themselves under a highway overpass, this is a bad idea. Winds in tornadoes only increase with height, so by going up towards the bridge deck, you are exposing yourself to stronger portions of the tornado. Not only that, by going into a narrow channel (between the bridge and the surface), you are moving into an area that will naturally produce stronger winds, thanks to something called the Bernoulli effect. It states that fluid will move faster the more restricted its channel is. If you can't reach a solid structure, the best thing to do is to find a ditch or a low-lying area and lie flat, while protecting your head as best you can.

Cold Air Funnel

Cold air funnel

Cold air funnel clouds are a rare type of funnel cloud that emanate from non-thunderstorm clouds; they often appear during partly cloudy skies, forming from towering cumulus clouds that aren't fully involved cumulonimbus. They form under many of the same circumstances as full-fledged tornadoes; in both, rotating air reaches out and attempts to reach the ground.

With that said, there are fundamental differences between cold air funnels and a traditional tornado. They aren't associated with dark wall clouds, mostly because they occur in situations with less cloud cover, and they aren't nearly as strong if they do reach the ground, and this is significantly rarer than with normal funnel clouds.

Staying Safe

Obviously, if you see any sort of funnel cloud, take cover immediately. Even though cold air funnels are weaker than normal tornadoes, they still are nothing to trifle with.

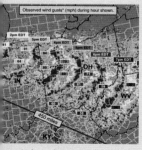

A derecho on radar

Derechos

Derechos are not identifiable by the naked eye, simply because they are typically quite vast. Derechos develop thanks to a storm with forward momentum and winds moving all in the same direction throughout the lower levels of the atmosphere; this allows the downdrafts created by rain to develop behind the thunderstorm. The colder downdrafts reinforce the momentum of the storm, allowing the derecho to sustain itself over long distances.

Derechos produce long-lasting and very strong straight line winds, sometimes with winds of 120 mph or greater; such storms can be 100 miles long and progress across many states over the course of a 12-hour period.

A downburst

Downbursts/Microbursts

As we've noted, downdrafts are caused when the weight of moisture in a thunderstorm overwhelms the upward strength of a storm's rising air. At times, the updraft will become suddenly overwhelmed, and the downdraft can become locally destructive.

Staying Safe

When it comes to heavy winds and hail—whatever the cause—stay indoors. When winds are strong, airborne debris is a real danger, so you need strong walls and a roof overhead.

A flash flood

Flash Floods and Seasonal Flooding

Heavy rain can come at any time of year (though it's less likely in the winter in the Midwest and in the summer on the West Coast), and it's not always a result of a strong thunderstorm. Slow-moving rainstorms, tropical features, or a series of thunderstorms that tracks over the same location can lead to flash flooding, the rapid flooding of low-lying areas.

Typically speaking, a flash flood causes rivers and creeks to spill over their banks, but it also can mean that low-lying areas, such as underpasses, valleys, or even roads can flood as well. Often when there is flash flooding, there is a current below the surface, as water continues toward lower ground.

Mudslides and landslides are another threat with heavy rain, particularly where terrain is hilly. Heavy rain can weaken certain types of soils or bedrock; when loosened enough, dirt, trees, and anything else at the surface slides downward. This sediment and debris can wash out roads or bury structures, and if there are existing floodwater currents, they are even more of a threat.

Seasonal flooding is different than flash flooding. It occurs as a result of heavy snow in winter followed by rapid melting in the spring. This can be made worse by an especially warm period at higher elevations or by a wet spring. Seasonal flooding often occurs in the same regions year over year, and this can be monitored as rivers rise with meltwater and runoff and the elevated river level moves downstream over time.

Staying Safe

Whatever the cause of floodwaters, it is important to stay away from them. It's impossible to tell how deep a flooded area is or how strong its current is just by looking at it. That's why it's always important to stay out of floodwaters, especially if you're in a vehicle. It only takes a few inches of water to move a car, and many flooding deaths occur when drivers try to pass through a flooded area only to discover that it's deeper than their vehicle is tall. Floods often wash out roadways too.

Flooding can often lead to evacuation orders. Seasonal river flooding leads to evacuations near the rivers' banks, but it tends to give affected residents a little bit of time to pack and move important belongings to higher levels. Flash flood-induced evacuations are much more urgent and often indicate mudslides or levy or dam failure; they could signal catastrophe on the way. If a flash flood evacuation is issued in your area, get out while you can.

Hyperthermia and Heat-related Illnesses (and Sunburn)

Beware high of temperatures

High temperatures and sun can cause their own problems. Sunburn affects everyone, regardless of skin color; the best prevention, of course, is applying sunscreen and seeking shade. If you haven't figured the best way to prevent sunburn, consult a doctor for advice.

Hyperthermia and heat-related illnesses are more-serious concerns. These conditions can be exasperated by humidity; symptoms include dizziness, confusion, and nausea. The heat can lead more directly to long-term kidney issues and can even be fatal in the short term, particularly among the elderly or small children.

As with most temperature-related ailments, there is no fixed temperature that is or is not dangerous. Exposure to direct sunlight can become dangerous at 75 degrees if it lasts for too long, and increased humidity can shorten the amount of time it is safe to be outside. Physical exertion can also accelerate the effects of the heat on the human body. The Heat Index is a calculation that tries to approximate what a temperature will feel like for the human body, and how dangerous a particular combination of temperature and dew point might be.

Staying Safe

Keep hydrated if you must stay outdoors, and make sure that you and your loved ones have adequate access to air-conditioning. Hyperthermia and heat-related illnesses don't just occur because of physical exertion; many fatalities are simply do to people being stuck indoors without a way to cool down.

NWS Heat Index

Relative Humidity (%) \ Temperature (°F)	80	82	84	86	88	90	92	94	96	98	100	102	104	106	108	110
40	80	81	83	85	88	91	94	97	101	105	109	114	119	124	130	136
45	80	82	84	87	89	93	96	100	104	109	114	119	124	130	137	
50	81	83	85	88	91	95	99	103	108	113	118	124	131	137		
55	81	84	86	89	93	97	101	106	112	117	124	130	137			
60	82	84	88	91	95	100	105	110	116	123	129	137				
65	82	85	89	93	98	103	108	114	121	128	136					
70	83	86	90	95	100	105	112	119	126	134						
75	84	88	92	97	103	109	116	124	132							
80	84	89	94	100	103	113	121	129								
85	85	90	96	102	110	117	126	135								
90	85	91	98	105	113	122	131									
95	86	93	100	108	117	127										
100	87	95	103	112	121	134										

Likelihood of Heat Disorders with Prolonged Exposure or Strenuous Activity

☐ Caution ☐ Extreme Caution ☐ Danger ■ Extreme Danger

Drifting snow

Blizzards

Not all dangerous weather happens in the summer. Blizzards are the strongest winter storm; the word blizzard is used colloquially to mean any bad winter storm, but a blizzard warning has a specific meaning, meteorolog-ically speaking. Blizzard warnings are issued for storms with heavy snow and winds of 35 mph or greater that last for three hours or more. These conditions usually bring white-out conditions or situations where visibility is near zero.

Despite how common the term is, the actual warning criteria aren't met as often as one thinks. Blizzards are seen most commonly in the northern Plains, where winds can blow unchecked by hills, trees, or other terrain; they're also seen in the strongest Nor'easter on the coast. While a city like Minneapolis is infamous for its winter weather, until its most recent blizzard in the spring of 2018, Minneapolis went more than a decade between blizzards. Blizzard warnings aren't handed out easily, and if you find one issued for your area, you should be prepared for a very serious winter storm.

Staying Safe

Snowy and icy weather produces perils on the roadways, while wind and cold temperatures can be dangerous to anyone caught outdoors.

Snow and ice create a greasy, slick surface that is tough for a car to navigate. Sometimes black ice forms; it's a layer of ice so thin that it can't be seen, so it looks black because the black pavement shows through. It can be created by dew, fog, or a very light rain freezing on a cold roadway, or, when it's particularly cold, by the moisture in car exhaust freezing onto the road. If the conditions are treacherous, there isn't much that one can do except to drive responsibly and slowly. Don't put yourself in a position where you have to change velocity quickly. Avoid accelerating or braking quickly and making abrupt changes in direction. Should things go wrong and you get stranded in your vehicle in a wintry scenario, it's important to have a safety kit, including blankets, flares, food, and perhaps a shovel or sand, for traction, if you are stuck in snow.

Hypothermia and Frostbite: If you get stuck outdoors in the cold without sufficient clothing, get indoors and dry off if you are wet, and drink a warm beverage, cover yourself in blankets, and change into dry clothes. Hypothermia starts with chills and general confusion and disorientation, and it can advance to cardiac arrest if left untreated.

Physical activity is good for mild cases of hypothermia. If a companion is experiencing an extreme case of hypothermia, with severe confusion, including a belief that they are overheating, seek immediate medical assistance. (Such paradoxical undressing is a sign of severe hypothermia.)

Frostbite is another danger of cold air, and it comes directly from exposure to cold air. Frostbite is a skin malady that is caused when the skin literally freezes. This can lead to blistering, numbness, and, in severe cases, the loss of fingers or toes. To avoid frostbite, ensure that no skin is exposed to extreme cold for very long, and try to rewarm soon after exposure to the extreme cold.

The threat that cold air poses on the human body is fairly well understood, particularly when it comes to frostbite. The wind chill is similar to the heat index, in that it combines temperature with another variable, in this case wind speed, to try to estimate the danger posed. The wind chill chart includes a "feels like" temperature, and a color code representing how long until the effects of frostbite might take hold on exposed skin.

Wind Chill Chart	Temperature (°F)																	
Calm	40	35	30	25	20	15	10	5	0	-5	-10	-15	-20	-25	-30	-35	-40	-45
5	36	31	25	19	13	7	1	-5	-11	-16	-22	-28	-34	-40	-46	-52	-57	-63
10	34	27	21	15	9	3	-4	-10	-16	-22	-28	-35	-41	-47	-53	-59	-66	-72
15	32	25	19	13	6	0	-7	-13	-19	-26	-32	-39	-45	-51	-58	-64	-71	-77
20	30	24	17	11	4	-2	-9	-15	-22	-29	-35	-42	-48	-55	-61	-68	-74	-81
25	29	23	16	3	9	-4	-11	-17	-24	-31	-37	-44	-51	-58	-64	-71	-78	-84
30	28	22	15	8	1	-5	-12	-19	-26	-33	-39	-46	-53	-60	-67	-73	-80	-87
35	28	21	14	7	0	-7	-14	-21	-27	-34	-41	-48	-55	-62	-69	-76	-82	-89
40	27	20	13	6	-1	-8	-15	-22	-29	-36	-43	-50	-57	-64	-71	-78	-84	-91
45	26	19	12	5	-2	-9	-16	-23	-30	-37	-44	-51	-58	-65	-72	-79	-86	-93
50	26	19	12	4	-3	-10	-17	-24	-31	-38	-45	-52	-60	-67	-74	-81	-88	-95
55	25	18	11	4	-3	-11	-18	-25	-32	-39	-46	-54	-61	-68	-75	-82	-89	-97
60	25	17	10	3	-4	-11	-19	-26	-33	-40	-48	-55	-62	-69	-76	-84	-91	-98

Wind (mph)

Frostbite Times ▢ 30 minutes ▨ 10 minutes ▣ 5 minutes

Wind Chill (°F) = 35.74 + 0.6215T - 35.75(V$^{0.16}$) + 0.4275(V$^{0.16}$)
Where, T = Air Temperature (°F)   V = Wind Speed (mph)

A 22-degree halo, sun dogs, and more

CLEAR-SKY PHENOMENA

Not all meteorological phenomena pertain to clouds or are associated with thunderstorms or severe weather. In fact, there are many meteorological features you might see even when the sky is relatively clear. These are known as clear-sky phenomena.

Most things that we see in the sky are a product of the interplay between the sun and the various forms of water or ice in the atmosphere; the amount of water/ice present, and the altitude that it occurs at, are the main variables in determining what phenomena we see. Let's take a look at some of the most notable phenomena that you might see if the skies *aren't* gray.

A rainbow

Rainbows

Rainbows are perhaps the most famous clear-sky phenomena. Famous for their beauty and ubiquitous in culture, they occur when light is reflected and refracted through water in the sky. They also serve as one of the best tools to explain how sunlight and the visible light spectrum work.

When we see the sun, it looks yellow, red, or orange. But when seen from space, the sun looks white. White light is a combination of all the different colors (wavelengths) on the visible light spectrum; to see that spectrum, you need a prism, which helps separate out the various wavelengths of white light. Organized by wavelength from longest to shortest, the visible light spectrum is ordered: red, orange, yellow, green, blue, indigo, and violet. (It should be no surprise that the next waves on the wave spectrum from visible light are infrared and ultraviolet, longer and shorter wavelengths respectively).

When this white light, which contains all the spectrum/colors in the rainbow, leaves the sun, it passes through outer space before hitting Earth's atmosphere. When the white sunlight bumps into molecules in the air, certain wavelengths, such as blue, are scattered more easily. This is why our sky looks blue; similarly, when we look toward the sun, we can see the wavelengths that don't scatter as much. This is why the sun appears yellowish and the sky appears reddish at sunrise and sunset.

Even though some light is scattered, nearly all of the available solar light reaches the Earth. Sometimes, sunlight bumps into water droplets instead of air molecules. When this happens, the light refracts (bends) inside the water droplet; it then reflects off the water droplet again, refracting again as it leaves. This process separates the white light into its component colors—a rainbow! The colors in a rainbow match the order of the visible light spectrum: red, orange, yellow, green, blue, indigo, and violet.

While we're familiar with the rainbow's bow shape, all rainbows actually create a full circle of a rainbow, but to see it, the conditions have to be right, and you have to spot the rainbow from a high elevation. Otherwise, half your view is blocked by the horizon. Usually, the best way to spot one is in an aircraft of some sort.

A double rainbow

Double Rainbows

Sometimes, when the light that creates a rainbow enters water molecules, it reflects more than once. When this happens, a double rainbow is produced; the normal rainbow is at the bottom, and a looser second rainbow is at the top. If you look carefully, you'll also notice that the top rainbow's colors are reversed.

Recently, triple and even quadruple rainbows have been spotted, but seeing them usually requires advanced equipment.

When and Where You'll See It: Because of the optics involved, the sun needs to be behind you in order for you to see a rainbow. Since most convection (heating in the atmosphere) occurs in the afternoon, when the sun is to the west, rainbows are most frequently seen to the east.

Green, Eerie Skies (and More)

Green skies

When thunderstorms arise, the skies darken; this occurs because the amount of moisture in the atmosphere filters out almost all of the sunlight shining through the sky, save for the longest wavelengths of dark blue and violet. Though there isn't much research on the topic, and the cause is unclear, when the clouds take on a green hue, severe weather may be on its way. One explanation may be the high moisture and ice content of the clouds, as well as the sun angle above those clouds.

Yellow Skies

Yellow sky

Sometimes sleet or freezing rain has the ability to turn the skies and clouds a shade of yellow; this occurs because the icier precipitation filters out a slightly shorter wavelength than the precipitation found in summer.

Reds and Pinks

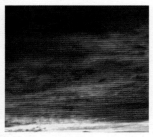

Red sky

Other common sky hues that we see, such as the reds, pinks, and purples of sunset, don't actually have anything to do with the clouds or rainfall in the area, but rather the curvature of the Earth, and the angle at which the sun's rays reach the surface. As the sun goes down, its rays are refracted through more of the atmosphere because it has a longer path through the atmosphere to our eyes; the other wavelengths of light are largely removed in the process, and we're treated to the brilliant shades of reds and pinks as the sun rises or falls. Particularly red skies are often caused by large amounts of particles high up in the atmosphere—such as those produced by wildfires or volcanic eruptions—in the atmosphere. Contrary to popular belief, however, run-of-the-mill air pollution and smog (which occur at lower levels) actually mute sunsets somewhat.

When and Where You'll See It: Green skies are likely to show up underneath showers and thunderstorms featuring very heavy rain, particularly in the afternoon and evening. Yellow skies are most likely in ice storms with somewhat thin clouds aloft. Reds, pinks, and oranges are enhanced when there are particulates in the higher levels.

Corona

Corona

The corona is the name of the outermost part of the sun's atmosphere. Consisting of a shell of hot gases, it is invisible in nearly all circumstances. The only time the corona can be seen on Earth is during a solar eclipse, when the moon slides in to block most of the sun's light. At that point, the corona can be seen from Earth as it peeks around the moon. **Important note:** Unless you're viewing the sun during a total eclipse—and only during the very brief period known as totality—do not look at the sun unaided, as eye damage may result. For details on spotting eclipses and safety, visit: https://eclipse.gsfc.nasa.gov/solar.html

Just to make things confusing, there is also an optical phenomenon called a corona, and this too can either involve the sun or the moon, but the atmospheric effect stems entirely from phenomena here on Earth. Coronas occur when a disc forms around the sun or the moon. This happens when light shining through the thin clouds (usually stratus or cirrus clouds) is diffused, or scattered through the cloud layer, instead of passing uninterrupted to the surface. This creates a disc of light around the sun or moon. Lunar coronas are more prominent because of the lack of ambient light, and they are most common when the moon is full or nearly full.

When and Where You'll See It: Lunar coronas are easiest to see when the moon is full and it's dark outside; solar coronas are harder to see because the diffuse light through a thin cloud is washed out by daylight. The sun's corona is difficult to see without specialized gear/safety equipment. (Reminder: Never stare directly at the sun without proper safety equipment.)

22-Degree Halo

A similar optical phenomenon to the corona is the 22-degree halo. It gets its name because the halo is found at an angle of 22 degrees from the center of the celestial body (whether the moon or the sun). Unlike coronas, the 22-degree halo looks something like a circular rainbow; these halos are caused by light refracted by ice crystals. The ice crystals that cause the halo

22-degree halo

sometimes stem from very high cirrus clouds, but it is just as likely that the ice crystals that created the halo aren't themselves visible.

When and Where You'll See It: Typically, the 22-degree halo is visible only in cold weather, when there are ice crystals available in the atmosphere to reflect and refract light.

Sun Dogs/Moon Dogs

Sun dogs and moon dogs are essentially the same phenomena; the only difference is whether they are seen around the sun or the moon. Sun dogs are more common, as the moon needs to be fairly bright for moon dogs to become visible. An atmospheric phenomena related to the 22-degree halo, a mock sun (or moon) is often found on one or both sides of the sun.

Sun dogs

Sun or moon dogs appear when light along a 22-degree halo (which isn't always visible itself) is refracted by ice crystals that are parallel to the Earth's surface. The larger the field of ice crystals, the larger the sun dog is. The refracted light of a sun dog will likely be red, though its color can move towards the blue end of the spectrum the farther from the sun it appears.

When and Where You'll See It: Like with the 22° halo, sun and moon dogs tend to appear most frequently during the coldest weather. They are also the most common when the sun is near the horizon.

Sun Pillars

Sun pillar

Sun pillars occur most frequently during sunrise or sunset, as the sun reflects off of ice crystals and high clouds on the eastern or western horizon, forming a column of light that extends vertically from the sun. Light pillars and moon pillars are also possible, seen above streetlights or the moon, for example. The causes for these are the same.

Northern Lights

Northern lights

The Northern Lights, or Aurora Borealis as it is known in Latin, is one of the most spectacular phenomena on Earth. Unlike many of the other phenomena we see here, it's not a product of sunlight interacting with ice or water in the atmosphere; instead, it's the result of highly energetic electrons from the sun interacting with molecules of the atmosphere itself.

The sun emits more forms of energy than just visible light; it produces waves along the entire electromagnetic spectrum. Some of those waves can interact with the very top of the atmosphere and ionize some molecules, particularly oxygen and nitrogen. Ionization is the process by which electrons are added or removed from an atom.

The process of ionization of those gases results in the colorful display we see at night, typically towards the poles. The Northern Lights are brought on by geomagnetic storms—disturbances to Earth's magnetic field caused by outbursts of energy from the sun. In that respect, they are space weather, and they don't portend any type of weather here. But all that geomagnetic activity can mess with electronic signals, such as cell phones or satellites.

When and Where You'll See It: The closer to the poles you are, the better. Generally speaking, for the Northern Lights to be visible, the sun has to be emitting geomagnetic energy, and you need to be in the northern portion of the country. Clear skies are also essential, as the lights occur at the highest levels of the atmosphere. For space weather forecasts, including details on whether the Northern Lights might be visible in your area, visit: www.swpc.noaa.gov

Noctilucent clouds

Noctilucent Clouds

Noctilucent clouds are high-altitude, whisper-thin clouds that exist at the top of the atmosphere. They are clouds belonging to the cirrus family but have a few important distinctions.

Noctilucent clouds are even higher than the average cirrus clouds, and because of that they contain far less actual moisture than typical cirrus clouds. The other important thing to note is that, most of the time, noctilucent clouds are invisible.

Noctilucent clouds only show up at sunset, when the angle of the sun across the atmosphere catches the ice crystals within them at just the right angle, leaving a thin wash of bright but thin clouds across the evening sky. Noctilucent clouds are most likely to occur in high latitudes.

When and Where You'll See It: You are most likely to see noctilucent clouds in the high latitudes and at twilight, when they catch the setting sun at just the right angle.

Nacreous clouds

Nacreous Clouds

Nacreous clouds are also known as polar stratospheric clouds, which should tell you that they appear mostly at the poles and at high elevations. Nacreous clouds tend to appear most vividly at twilight and shine in a wide array of iridescent colors. This is not to be confused with iridescent clouds, which look superficially similar, but occur when individual clouds have a rainbow- or soap bubble-like coloration due to the refraction of water droplets or ice crystals.

Because of where they appear, the polar stratosphere, nacreous clouds make use of the coldest observable space in the Earth's atmosphere. Water doesn't necessarily even exist at those altitudes, let alone condense into clouds. Instead, nacreous clouds form out of other chemicals, such as nitric acid or sulfuric acid, that are finally cool enough to condense, and their different chemical component accounts for the various color schemes that nacreous clouds can produce.

When and Where You'll See It: Nacreous clouds are most commonly seen near the poles and at twilight. Because of this, they are most widespread in the winter.

Fallstreaks

A fallstreak is a phenomenon that appears in high sheet clouds, like altostratus or cirrostratus clouds. A fallstreak appears as a hole in the blanket of clouds, created when supercooled water droplets freeze, causing other supercooled droplets around the initial ice crystal to freeze and start to fall, often forming a circular gap in the cloud layer.

A fallstreak

Lenticular clouds

Lenticular clouds are disc- or saucer-like clouds that form as air tries to flow around objects that are thrust into the sky, like hills, mountains, or buildings. The clouds form in the eddies created as air tries to move around the obstacles to the flow.

Lenticular cloud

Horseshoe Clouds

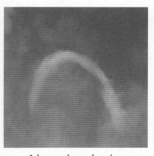

Horseshoe clouds are U-shaped spits of clouds that are usually ejected from larger cumulus clouds. They are caused by updrafts catching crosswinds aloft, with stronger winds pushing part of the clouds faster than the portion in the middle, bending the cloud into a horseshoe shape.

A horseshoe cloud

Asperitas Clouds

Asperitas cloud

Asperitas clouds were recognized as their own type of cloud formation only within the last decade. Asperitas clouds have wavy bases and can blanket the sky, often occurring near storm clouds, though they don't forebode precipitation.

FIND YOUR WEATHER, WEATHER SAFETY, AND MORE

The media is one of the greatest tools for ensuring weather safety. Broadcast television stations relay National Weather Service weather watches and warnings, particularly for severe thunderstorms, tornadoes, and flash floods, while weather radio is also a crucial link when it comes to providing the public with severe weather updates.

Unfortunately, there is a great deal of confusion over the terminology surrounding weather watches, warnings, and advisories. Given how complicated our weather terminology is nationwide, perhaps this section won't change that confusion across the country, but hopefully this section will make things clearer for you.

Watches and Warnings Apply to Specific Geographic Areas

One of the greatest issues with broadcasting watches and warnings is a problem of geography: people often don't know how to pinpoint their own location on the map, especially if they're traveling. Since you could be reading this book anywhere, I'll leave it to you to figure that out. Watches, warnings, and advisories are almost always issued with reference to counties (or parishes/boroughs, depending on the state jurisdictions), so it is important to know the name of the county you find yourself in, above all else. In states with many different counties—Minnesota has 87—it's important to keep track of your location so you can know which watches/warnings apply to where you are.

NOAA weather watches

Weather Watches

After you have familiarized yourself with your local geography, you need to know the difference between a weather watch and a weather warning.

A weather watch, such as one for a severe thunderstorms or tornadoes, means that conditions are favorable for the development of the extreme weather in question. For example, if you're in an area with a severe thunderstorm watch, a severe thunderstorm may not be present in the area, but conditions are right for one to form. As a result, a fairly large area is being monitored for the development of those storms and for an extended period of time.

A Weather Watch Example: A typical weather watch, for example, one issued at 2 p.m. in Minnesota, would say something like, "A Tornado Watch is in effect within 100 NM (nautical miles) of a line from 20 miles northwest of St. Cloud to 15 miles west of Worthington until 11 p.m. This Tornado Watch is in effect for the following counties…"

The counties affected would then be listed off in alphabetical order, as would the threats that tornadoes pose. With a weather watch, it is important to stay tuned for weather events, but there is no imminent or immediate danger.

Weather Warnings

When there is a weather warning, it means that severe weather is occurring and expected to arrive soon in the warned location, if it is not there already. Weather warnings focus more directly on a smaller location and last for a shorter amount of time.

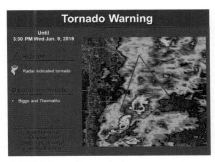

A NOAA tornado warning

A Weather Warning Example: Again, an example is helpful. A tornado warning would read something like the following, "A tornado warning has been issued for Kandiyohi County, Minnesota, and western Meeker County, Minnesota. At 4:42 p.m., Doppler Radar indicated a tornado 3 miles east of Raymond, or 10 miles southwest of Willmar. This possible tornado is moving east-northeast at 30 mph. It is expected to be near Svea at 4:46, Rosendale at 4:52, Litchfield at 4:59. This warning is in effect until 5:25 p.m." The station would then list the threat that tornadoes pose and advise residents on the best course of action.

The Weather Conditions that Trigger a Warning

A **severe thunderstorm** is defined as a storm with winds of 58 mph or greater or hail an inch in diameter or larger.

A **tornado** warning is issued if a tornado or funnel cloud has been observed or if radar indicates the tight rotation of a potential tornado.

Flash flood warnings are issued when heavy rain is expected, but there are no warnings designated for lightning.

Blizzard warnings are issued for storms with heavy snow and winds of 35 mph or greater lasting for three hours or more. These conditions usually bring white-out conditions or situations where visibilities are near zero.

Winter storm warnings can be issued for a wide variety of winter weather concerns, and the criteria for issuing them varies from place to place. A winter storm warning will be issued if impactful snow or ice is falling on an area, with the local National Weather Service determining exactly what might be impactful.

It is important to note that lightning and heavy rainfall are not elements of a severe thunderstorm warning, despite the dangers that they may pose.

Tornadoes: Observed or Radar Indicated?

When it comes to tornado warnings, the National Weather Service refers to tornadoes either as observed or radar indicated.

Generally speaking, both of these descriptions are equally reliable (or unreliable, in some cases). Often, the observed tornadoes are funnels that aren't yet on the ground or are optical illusions of the night. Radar-indicated tornadoes are measured aloft, and they suggest an environment that is strongly favorable for the development of a tornado, but not the confirmed presence of a tornado. While these descriptors are not perfect, they both warrant taking cover as soon as possible.

If the tornado is observed on the ground by a storm spotter or emergency personnel, it is labeled as a confirmed tornado. A **tornado emergency** can be issued if a confirmed tornado is moving towards a populated area. This is the most concerning and dangerous type of severe weather warning, and it should be acted upon urgently.

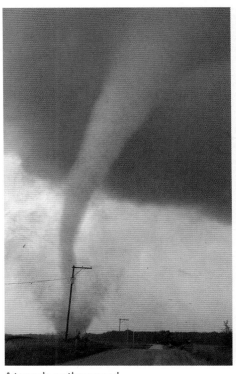

A tornado on the ground

Winter weather

Weather Advisories

Of course, severe thunderstorms and tornadoes aren't the only types of weather that can impact our lives or our livelihoods, so the National Weather Service has another alert criteria, the advisory, which is most often seen in the winter, though advisories are often issued in the summer as well.

Advisories follow the same course of logic as a warning. They are issued with the certainty of a warning—when forecasters are sure that the type of weather phenomena is ongoing or will be underway soon, but when the weather event in question isn't as severe as other events that call for a warning.

The best and most frequent example of an advisory is a Winter Weather Advisory, which is issued when snow, blowing snow, or other wintry precipitation is expected. Advisories can be issued with different criteria, depending on the location. It might take a few inches of snow for an advisory to be issued in Minneapolis, but a trace of snow might be all it takes in Atlanta for an advisory to be issued.

In the summer, similar advisories are issued for strong, but not severe, thunderstorms. Other types of advisories are issued for individual types of phenomenon, like wind, heat, or fog.

Find Your Local Weather

The greatest tool that the modern meteorologist, weather enthusiast, or outdoorsperson has today is the internet.

The U.S. government has better weather information and tools than perhaps any other organization in the country. In fact, the primary weather models and imagery that forecasters use to predict the weather are effectively products of the U.S. government (or derived from them). Better yet, they are available for free to the general public.

Stay informed

The National Weather Service

The main face of the government's weather endeavors is the National Weather Service, charged with providing the public with forecasts, observations, and warnings. Their website, www.weather.gov, is an excellent resource for weather information, and it provides a wide variety of radar and satellite imagery, as well as a comprehensive list of products

Part of the National Weather Service's Nexrad Doppler Radar Network

pertaining to flooding and hydrology. The best resources it provides, of course, are the daily forecasts.

The National Weather Service page provides forecasts for any point in America, stretching out a week in advance. You can punch your locality's name into the search bar, or you can visit your local Weather Service office's page by clicking on the map of the United States on the homepage. For a more robust discussion of the weather forecast, use the "Forecasts" menu and select "Forecast Discussion."

The Storm Prediction Center

If you want to learn of the threat for thunderstorms and severe weather in your area, visit the Storm Prediction Center (www.spc.noaa.gov). The SPC issues all tornado and severe thunderstorm watches (local

The convective outlook from the Storm Prediction Center

National Weather Service offices issue the warnings), and their site contains graphics and information pertaining to such watches. Additionally, the SPC produces fire weather prediction information and provides the details on any severe storms for the day, as well as those reports archived through history.

The Convective Outlook

One of the Storm Prediction Center's most useful tools is the Convective Outlook. This is a graphical look at where thunderstorms might be expected in a day, as well as an initial look at where thunderstorms might be severe and where the threat for the worst thunderstorms may be. The SPC has several categories for forecasting whether severe thunderstorms will affect an area: marginal, slight, enhanced, moderate, and high. On

An example forecast from the Storm Prediction Center

a day with a slight risk, severe weather might be expected somewhere in the region; on days with a moderate or high risk in an area, a severe weather outbreak is essentially expected. The outlooks are issued daily and can extend out up to eight days, and they are a great first stop for those concerned with thunderstorms.

Precipitation Prediction

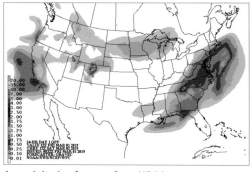

If you are looking strictly for precipitation forecasts, the Weather Prediction Center is the site for you (www.wpc.ncep.noaa. gov). The WPC provides an overview of the weather forecast on its main page,

A precipitation forecast from NOAA

as well as links to forecast maps for excessive rain or winter weather, most notably heavy snow and ice. Specifically, there is a link to QPF, or quantitative precipitation forecasts; these produce a color-coded map showing the forecast total amount of liquid precipitation over the designated time period.

The National Hurricane Center

If you live or plan to travel along a coast, particularly the East or Gulf Coast, the National Hurricane Center (www.nhc.noaa.gov) is indispensable, particularly in late summer or early autumn, when hurricanes are most frequent. The NHC issues tropical outlooks for the Atlantic and eastern Pacific (as far west as the American territories of Guam and the Mariana Islands).

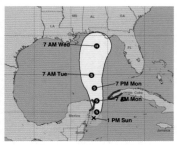

A forecast from the Storm Prediction Center

When there are tropical features in either ocean, they are highlighted on a satellite image on the homepage, and a suite of new forecast products are displayed on the site, including a discussion of the current position and strength of the feature, as well as a forecast for where the system might go, with a probability cone highlighting the possible range of the track that a tropical storm or hurricane might take. Additionally, there will be color-coded forecasts for expected wind strength, and a display of the various tropical storm and hurricane watches and warnings on threatened coastal regions.

A Weather Bureau Forecast from 1911; forecasts have come a long way.

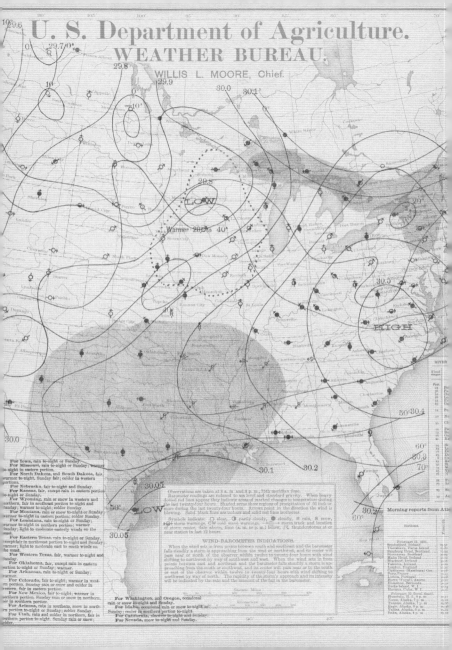

U. S. Department of Agriculture.
WEATHER BUREAU.

WILLIS L. MOORE, Chief.

Setting Up a Home Weather Station

One of the easiest ways to get involved with the weather is to set up your own weather station. A weather station manufactured by AcuRite is pictured here, but there are many other whole stations or components that you can buy and set up. The model seen at left is associated with an electronic display that is kept indoors, though there are many weather monitoring devices that you can see from a distance and that won't tax your wireless network when collecting observations. Most weather stations collect much of the same types of data.

Rain Gauges

Most people start out in backyard meteorology by simply putting out a basic rain gauge. Low-tech models consist of a cylinder held upright with measurements clearly marked on the outside; these need to be emptied after rain events so you can keep an accurate record of the next rainfall event. More-sophisticated systems include a pass-through rain gauge, which allows rain water to move past a sensor that keeps tabs on the amount of water, and there's no need to empty it.

An ornamental rain gauge

Anemometers, Wind Vanes, and Wind Socks

An anemometer is a device with cups that catches wind from any direction and tracks wind speed. The cups spin with the wind, and the rotations of the device measure the speed. A wind vane is a lower-tech wind-monitoring device (and often a decorative piece of yard art); it shows the direction of the wind and points in the direction that the wind is coming from. Commonly seen at small

An anemometer

airports, wind socks, which fill with air and point in the direction the wind is blowing, are another option too.

A backyard thermometer

Thermometers

If you want to track temperature, be sure to purchase a thermometer built with meteorology in mind. Such devices take positioning (relative to direct sunlight) into account. A thermometer purchased at the local garden center will not, and it will give you different temperatures if you hang it from a tree or if it is posted somewhere in direct sunlight. Don't make any weather-related decisions based on such devices, as they will prove inaccurate.

Barometers

Barometers are widely available. They are used to measure atmospheric pressure, which is now recorded in millibars, instead of in inches of displaced mercury as it once was. There are all manner of barometers for public consumption. There is no real trick for placing barometers,

A vintage barometer

so long as they remain upright and outdoors where they can measure the weight of the atmosphere on top of them. Placing a barometer indoors doesn't provide much meteorological benefit.

A Few Devices You'll Only Find in Automated Weather Stations

There are a few devices you will usually only find in an automated sensor like the one seen on page 118. A hygrometer, which measures humidity, and a pyranometer, which tracks solar radiation, are the most common examples. Other examples of specialized devices include transmissometers, which measure visibility, and ceilometers, for measuring cloud heights. (Both are quite common in the aviation industry, as you might expect.)

Where to Set Up an Automated Station

The best place to put an automated station like the one on page 118, particularly one with the anemometer and wind vane, is at an elevated position that is not obstructed in any direction. Most self-contained weather stations have internal hygrometers, thermometers, and barometers, so it is more important to not have overhead obstructions so rain gauges can collect water properly. If the weather station has separate components, particularly an external thermometer, more accurate readings can be taken if the thermometer is placed in the shade.

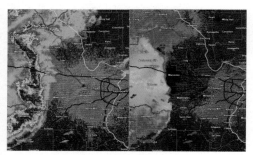

A bow echo on NOAA radar

Become a Weather Spotter

If you're interested in the weather, becoming a weather spotter is a great way to learn more about weather while helping ensure public safety. Your local National Weather Service office is always looking for weather spotters. Spotters help report weather observations, particularly severe weather, at locations far afield from the weather service office, airports, or municipal observation stations. This helps offer a more complete picture of the weather picture.

Spotters go through a brief training seminar, often at their local National Weather Service office. There, they get a more complete review of observable severe weather and the observations that are particularly helpful when advising the public. I would encourage anyone with an interest in the weather, especially those who find themselves in remote locations often, to consider volunteering to be a spotter.

Do-It-Yourself Forecasting

Perhaps you have had enough of the local weatherman, and would like to see for yourself what the weather is going to do. Thanks to the dawn of the information age, and all of the weather information at your fingertips, it's never hard to get started!

Weather Models

Of course, before you look at any of the following resources, always defer to the professionals when severe or hazardous weather is predicted. Still, the following resources will help you see what meteorologists and forecasters are looking at when they create their forecasts or issue their warnings.

The National Centers for Environmental Prediction (NCEP) produce the **Global Forecast System** (GFS) and **North American Mesoscale Forecast System** (NAM), two of the most commonly used models by American meteorologists. The National Center for Atmospheric Research (NCAR) hosts those models at http://weather.rap.ucar.edu /model/ as well as other models that are of less renown.

As was noted above, the GFS and NAM are the two primary forecast models that meteorologists use, and they have their own particular advantages and disadvantages. The GFS models the entire planet, which means it's better at predicting global weather patterns. Because of this, it is far better at predicting longer-range forecasts. The NAM takes a narrower look, only calculating the forecast for North America. This means it can take a higher-resolution look at the weather across the country, but since it is unable to look at the global patterns, the NAM's forecasts have a shorter shelf-life than those of the GFS.

On the NCAR site, the precipitation forecasts are likely the most useful for the layperson, but the MSLP (mean sea level pressure) is a hidden gem, showing where the big systems are and where they might be heading. I recommend selecting the option to loop those models to get a full picture of how the features on the map are moving.

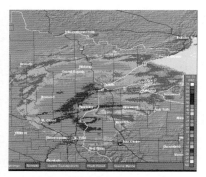

National Weather Service radar

Radar Programs

If you watch the local news, there are a couple of screens you are sure to see. The weekly forecast (with the sun wearing a pair of shades on a sunny day) and the radar. The radar is an incredibly important and useful resource for tracking ongoing weather, and it is one of the more captivating visuals when monitoring the weather remotely. Radar is readily available on most media outlets, but a more reliable version is available on the National Weather Service home page.

Radar is a fairly easy concept on its face. Keep an eye out for the greens, yellows, and reds on the radar image, as they represent precipitation in graduating degrees of intensity; if you do that, you will have a good idea of when to expect rain or snow.

On a standard radar, particularly the display offered by the National Weather Service, the scale goes from light to dark green, and then on to yellow, orange, red, purple, and white as more and more of the radar beam gets reflected back to the transmitter. Usually, rain isn't enough to generate those purples or whites, and when those colors show up, it's a good indicator of larger items falling from the sky, like hail.

But before you start looking at a radar map, however, confirm that you can locate where you are on the map; Google Maps is helpful for this.

Hook Echoes: Of course, there is more to a storm than the rain, and there is more to radar than whether or not it is raining. Perhaps you have heard the term hook echo when tornadoes threaten. The echo part of the terminology refers to the radar return, while the hook demonstrates a hook-like shape on the radar display.

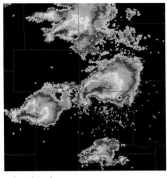

A hook echo

Supercell thunderstorms are rotating thunderstorms that often exist separate from other showers or storm activity. When a supercell produces a tornado, a hook echo shows up, usually at the southwestern end of a supercell. It is caused by the strong rotation of the tornado redirecting not only the air, but the rain that falls around it.

Hooks are most evident in discrete supercells, but they can also appear in embedded supercells, which are supercells that have merged with broader areas of showers and thunderstorms, though they may not be as prominent. It's important to note that tornadoes can occur even without an evident hook echo, so it is important to heed warnings for your area, even if you haven't identified a hook yourself.

Bow Echoes: A bow echo is a feature reflective of strong straight-line winds. It appears as an arc, pushed forward in the direction of the strongest wind flow. The momentum of continuous thunderstorm downdrafts propels the center of a line forward faster than the ends of the line, where the dynamics of the storm aren't as strong. On a radar screen, it looks like a bow shape.

A bow echo

Snowstorms: On radar, snowstorms are generally larger and fairly amorphous when compared to their rainy counterparts. If you're looking for distinct characteristics, it can be fairly difficult to figure anything out when the entire radar screen is bathed in light green. Often, however, there is one long stripe, perhaps no more than 5 miles wide (but perhaps dozens of miles in length) of a darker green or even yellow. This is called a bright band and signifies the area with the most intense snow. It doesn't move much, and that can lead the area below this stripe to accumulate a few more inches of snow than areas on either side of the band. It is fairly common to see more than one bright band in a snowstorm.

Privately Produced Radar Products

When it comes to radar, private products are also available. More information is available on some of the privately provided products mentioned below. They cost money, but they offer better resolution and a suite of options to dig deeper into storm activity. Some of those options include information about cloud heights (indicating the strength of the storm updrafts), vertically indicated liquid (a good measure for the threat of hail), and base velocity, which shows the direction and speed at which winds are blowing and can be used to identify either severe thunderstorms or tornadoes, if winds change direction in a tight area.

There are many examples of products available for purchase, and I recommend first and foremost that you shop around to find something that suits you, but I can share a couple of favorites. A desktop-based product from Gibson Ridge called GRLevel3 is clean, easy to use, and has all the data you could need. For mobile devices, RadarScope is one of the most popular and highly rated products available.

Recommendations

Below are some items and tools that I personally make use of that I would suggest to a hobbyist interested in exploring the weather more closely. They can be accessed or purchased by anyone.

Weather Station—I use and reference AcuRite weather stations, which need only Wi-fi (to monitor conditions) and a good place to be mounted. Explore the full selection at www.acurite.com

Radar—For desktop, I personally enjoy the Gibson Ridge display of radar products. I use GRLevel3 for my at-home radar display, though there are other varieties that you can explore on the Gibson Ridge page. The link here is for GRLevel3. www.grlevelx.com/grlevel3_2/

Weather Forecasting in a Pinch

Of course, you might find yourself out in the field, away from modern technology. If that's the case, the most important question you might ask is, "How can I tell if it's going to rain?" Unfortunately, there is no simple trick to figure that out. With that said, here are a few things you can do to reasonably divine some information about the weather in your area, just by making some observations.

1. Assuming there isn't obviously a storm bearing down on you (dark clouds, increasing winds, etc.), identify the direction the wind is coming from. Stand with your back to the wind, and point straight to the left. In doing so, you are pointing in the direction of low pressure. If you are finding yourself pointing to the west, northwest, or slightly to the southwest, the threat of rain is increasing.

2. Dark clouds seem like an obvious cue that inclement weather is approaching, but that fact is important enough to reiterate. The darker the cloud, the more of sun is being blotted out. In the summer, that is a sure sign that those clouds are laden with moisture. If you note that the clouds are dark and the wind is blowing towards the storm, be careful because that suggests localized low pressure, strong updrafts, and the threat for hail or even tornadoes, in conjunction with the already expected heavy rain. If wind is blowing from the storm towards your direction, the storm is likely collapsing, and the outflow is caused by the weight of falling rain suppressing the updraft and dispersing all that energy. The exception is when storms are strong and linear. In those cases, very strong winds will precede the advance of the thunderstorm, moving in the direction of the storm's motion, though wind is likely to be drawn into the storm before it turns around. Be wary of the so-called "calm before the storm," which is likely to precede the leading edge of thunderstorm-generated winds.

3. Clouds shift and change all the time. The most informative time for monitoring cloud changes is when you can see the cloud tops. If they are growing and building upwards, be prepared for some very nasty weather. The taller they get, the more likely that hail is in the future. Generally speaking, rapid cloud growth is indicative of instability in the area, and that indicates that thunderstorm activity is possible and could erupt rapidly.

Dark clouds in the distance

Weather Lore

There are many sayings and local traditions that proclaim to explain the weather, and even to predict it with unfailing accuracy. As you might imagine, most of these claims are not exactly backed up with evidence. Let's look at some of the most common and how reliable they are.

"Red sky at night, sailor's delight. Red sky in the morning, sailor take warning."—Fairly reliable! The weather generally moves from west to east, while the sun rises in the east and sets in the west. The red sky in the morning is caused because the rising sun's light is being scattered through clear skies, with clouds on the opposite horizon. If the sun is rising in the east, then eastward-moving clouds from the west are going to be overhead through the day. If the red sky is at night, the clouds are already to the east and ready to move out of the picture.

Red sky at night

"Halo around the sun or moon, rain or snow soon."—Somewhat reliable! The increasing moisture at high altitudes, which causes the halos, can foretell advancing low pressure, which may or may not mean precipitation.

"Rain before seven quits before eleven."—Not reliable. Weather systems come in all shapes and sizes and at all times of day. There isn't really a foundation for this particular saying.

"If the groundhog sees its shadow on February 2, there will be six more weeks of winter."—Not reliable. I'm sorry to tell the people of Punxsutawney that rodents are not reputable meteorologists.

"Lightning never strikes the same place twice."—Not reliable. Electricity takes the easiest path to complete a circuit. Often with lightning, that means connecting with a high point or finding a more conductive material at the surface. If anything, if a place is struck by lightning, it is probably slightly more likely to be struck again.

"Count the seconds between lightning flashes and thunder booms to tell how far away the storm is."—Somewhat reliable! There is definitely a relationship between when you see lightning and the length of time until you hear the thunder. Unfortunately, you can only really tell how close it is relative to the last strike you heard in the same thunderstorm. Humidity levels can affect the speed of sound, so a particularly humid day will make thunder sound closer than days that are less humid. So, yes, you can tell if a storm is getting closer by listening to the thunder after a lightning strike, but you can't really tell how close it is.

WEATHER CONVERSATIONS

It doesn't seem like there should be any controversy when it comes to the weather. Still, it is a topic that, given the right tone or the wrong conversation partner, can turn a bit adversarial. In the next section, we will touch upon a couple of the more hot-button issues that have their roots in politics or the media, two fields that always come with their own controversy.

The aftermath of Superstorm Sandy

Global Warming/Climate Change

There is probably no more quarrelsome topic in the country right now, perhaps in the world, than global warming or climate change. Unfortunately, the debate is not one that's had in good faith. Anthropogenic (man-made) global warming is a real issue, and is, for all intents and purposes, settled science. Those that deny this fact either attribute flawed science or got lost in the weeds of a political debate about the environment and our policies toward it that is now more than a generation old.

There has always been debate about the environment and to what extent we have a responsibility protect it, whether because of pollution, habitat fragmentation, or simple overuse of the land. The debate over global warming and climate change is a subset of that debate, and the solutions now being offered (e.g., switch entirely to renewables, enact a gas tax, turn to geoengineering) are a legitimate open debate. But instead of addressing those, some segments of the population call into question the settled science. So first, let's lay out what is going on with global warming and climate change so we can appreciate those debates a little bit more fairly.

Weather vs. Climate

Earlier, we discussed convection and compared it to a boiling pot of water. I like to use that example when comparing meteorology to climatology. Meteorology is akin to figuring out where the next bubble will appear in your pot of water. Climatology is asking if the stove is on.

Where meteorology concerns itself with individual weather phenomenon, particularly over the short term, climatology is the study of long-term weather patterns that characterize a particular geographic region, as small as the California coast to as large as the planet.

The Greenhouse Effect

The Earth's atmosphere retains heat thanks to the greenhouse effect. Solar radiation arrives in the atmosphere and is trapped by certain gases before it can be reflected back out into space. Carbon dioxide is one of those gases that traps heat in the atmosphere.

Human activity has upset the naturally existing carbon cycle. Today, the amount of carbon dioxide in the atmosphere is increasing. This added CO_2 in turn traps more of the sun's heat in the atmosphere. The sources of this carbon dioxide range from coal-fired power plants and cars to deforestation (plants absorb CO_2). CO_2 in the atmosphere is increasing exponentially, something that

Carbon dioxide on the rise

is monitored in real time, but also can be compared to previous time periods via ice and sediment cores and carbon dating.

Changing Terminology

Over the past decade or so, one thing that has changed in the discussion on global warming is the introduction of the term climate change. This can be confusing, but it's an important distinction: the planet's average temperature is going up, but the change in temperature is felt disproportionately at the poles. That doesn't mean that the rest of the planet doesn't feel the changes, though.

The phrase climate change refers to how Earth responds. The rising temperature changes the thermodynamics of the planet, and as a result, the atmosphere changes in an attempt to reach equilibrium. This has many concerning effects: the jet stream increases in amplitude, and this in turn allows for the development of stronger storms in all seasons, as well as more-persistent droughts, floods, heatwaves, and even some cold spells in temperate locations. (Yes, you read that correctly: An overall rise in the temperature of Earth's climate will disrupt the weather pattern enough to bring colder weather to some regions, at least temporarily.)

Future Impacts

Because the science around climate change is settled, the question is "What can we do now?" The fundamental cause of global warming and climate change is that the atmosphere has too much CO_2. We therefore need to reduce the amount of CO_2 we produce and find ways to help absorb the CO_2 already in the atmosphere.

What You Can Do

Reducing CO_2 emissions largely means reducing our reliance on fossil fuels. Conserving electricity, being more conscious of our driving decisions, and reducing air travel are all good ways to do that. Switching to sustainable power resources also necessarily reduces the use of fossil fuels.

To reduce the CO_2 that is existing in the atmosphere, one must look at the natural means by which it is absorbed. You might remember from your high school botany class that plants absorb CO_2 and exhale oxygen as part of the process of photosynthesis. Unfortunately, deforestation has weakened the planet's ability to reduce CO_2, particularly in the rainforests.

Trees are a natural carbon sink.

Planting trees, especially hardwood forests, can help reduce the existing CO_2 in the atmosphere and can also help get the carbon cycle closer to its natural balance.

The Challenge

Even if these changes were adopted on a large scale, the Earth is likely going to continue warming for the foreseeable future; the goal now is to slow the gradual rise in the Earth's temperature before it reaches what climatologists determine to be catastrophic levels (around 1.5-2 degrees Celsius or 3.6 degrees Fahrenheit).

Since the Earth is being heated disproportionately poleward, polar ice is expected to continue to melt and sea levels will rise, and with it the threat for some climate change, thanks to shifting oceanic currents. The best way to adapt to the inevitable changes is to manage the geology around us. Wetlands and watersheds are the best tools for managing water intake, and things like flash flooding can be allayed if construction in coastal wetlands is stopped. Concrete, on the other hand, is impervious to water and can make flooding worse.

In Defense of (Sort of) the TV Meteorologist

Of course, the news media is the primary method by which almost all of us get our weather information. I'm a meteorologist, so perhaps I am oversensitive to the opinion, but I'd guess that there are few professionals who are trusted less than the TV meteorologist.

While television forecasters certainly, at times, are guilty of hyping weather to attract viewers, many of the critiques leveled at TV forecasters are simply occupational hazards. All meteorologists face similar issues, but most forecasters don't have their hits and misses subjected to a wide audience.

Precipitation forecasts probably get the most attention on a daily basis, and they are usually issued as a percentage. A forecast calls for a 40 percent chance of rain, for example. But what that means can be something of a mystery. The first thing to consider is that a forecast isn't issued simply for your location, but for a wider region. Secondly, the mechanisms that produce showers and thunderstorms aren't necessarily efficient in producing precipitation: a storm might inundate a region in one case, whereas in another, a trigger for convection fails to materialize. Lastly, there is another issue: convection (which produces rainfall) can form in two different ways. Sometimes showers and storms spring up in isolation, whereas others form a broad swath of shower activity.

The forecast precipitation percentage, therefore, considers these three points: the variation across a forecast territory, the likelihood of precipitation across the region, and the type of precipitation and storm structure that is to be produced.

Consider a Snow Forecast

A perfect reflection of the challenges of forecasting for a region comes with snow forecasts. As we noted with the radar returns, narrow bands of heavy precipitation are common with snow events, which can make targeted forecasts difficult. Forecasts are often given with an eye towards accounting for those heavy slices of the region. "Snowfall accumulating to 2–4" with some spots seeing 6"" is a good example of that phrasing, though you may see forecasters simply say that snow of 2–6" is possible.

Still, the cutoff between heavy areas of snow can be dramatic, and it can make poorly phrased forecasts look foolish, when all possibilities aren't accounted for. Just know that when a forecast seems ambiguous, it's likely a reflection of the range of snow totals expected in a given location.

Some Troublesome Phrases

Another problem is the language used in forecasting. Partly cloudy and partly sunny, for example, mean the same thing, but when cloudiness is referenced in the forecast, it has a negative connotation. Another example occurs in precipitation forecasts. If there is a 20 percent chance for rain, the National Weather Service will mention the threat of rain in its forecast, as will most outlets. Some outlets, however, will wait until there is a higher chance for rain before mentioning it; that 20 percent chance is the most common threshold, but it still pays to find the percentage-based forecast, if you can, because it may reveal something beyond the text forecast of "chance of showers."

Mentioning the Worst

On the other hand, because media companies broadcast to a large territory, they often describe the forecast for a large region. That makes the job of forecasting much harder, as they need to mention, at the very least, the worst-case scenario, so those with the potential to be affected have time to prepare. If a severe thunderstorm is headed toward Dallas and Fort Worth, forecasters raise the alarm for both. If the worst case is achieved in Dallas, but Fort Worth doesn't see it, it often seems like the meteorologists are crying wolf, when they're really being prudent.

United States Weather Records
(All records considered only for the Continental United States)

All-time coldest temperature: -70°F (Lincoln, Montana, 1954)

All-time hottest temperature: 134°F (Furnace Creek, California, 1913)

All-time 24-hour rainfall: 42" (Alvin, Texas, 1979)

All-time 1-hour rainfall: 12" (Holt, Missouri, 1947)

All-time 1-minute rainfall: 1.23" (Unionville, Maryland, 1956)

All-time 24-hour snowfall: 75.8" (Silver Lake, Colorado, 1921)

Strongest wind recorded: 231 mph (Mount Washington, New Hampshire, 1934)

Most rapid change in temperature: 49 degrees in 2 minutes (Spearfish, South Dakota, 1943)

Greatest 24-hour temperature change: 100 degrees (Browning, Montana, 1916)

Longest period without rain: 767 days (Bagdad, California, 1912-1914)

Largest hailstone: 7" in diameter (Aurora, Nebraska, 2003)

Recommended Reading

Holton, James R. *Introduction to Dynamic Meteorology.*
   If you ever want to get really deep into the science and math of
   meteorology, get this book. I haven't met a meteorologist who didn't
   have a textbook from Holton.

National Weather Service. *Weather Spotter's Field Guide*
   This online PDF provides details about severe weather for the purpose
   of storm spotting. (Available here: www.weather.gov/media/bis
   /Weather_Spotter_Field_Guide.pdf)

Laskin, David. *The Children's Blizzard.*
   This look at a catastrophic blizzard in the nineteenth century not
   only delves into the impacts that a severe winter storm can have,
   but it also offers insight into early weather forecasting techniques and
   their effectiveness.

Glossary

Alberta Clipper A common North American storm system that emerges from the Canadian Rockies

Barometric Pressure A measure of the weight of air on top of a certain point

Blizzard Winter storm with heavy snow and winds of 35 mph or greater lasting for three hours or longer

Bomb Cyclone A rapidly developing storm system that sees its central pressure drop by 24 mb in 24 hours

Bow Echo A series of thunderstorms with straight-line winds that appears as an arc on radar, pointed in the direction the storm is moving

Cirrus Wispy clouds found extremely high in the sky

Cold Front The leading edge of an advancing mass of relatively cold air

Cumulonimbus Cumulus clouds that have become stronger and are capable of producing rain, snow, or thunderstorms

Cumulus Puffy clouds that can occur at low or middle levels of the sky

Derechos Long-lasting straight-line windstorms that can stretch over 100 miles and cover many states over the course of a 12-hour period

Dew Point The temperature at which moisture in the air can begin to condense into water droplets

Downdraft An air current caused by the weight of falling precipitation, overwhelming an updraft

Flash Flood The rapid flooding of low-lying areas

Fog Low stratus cloud that has reached the ground or emanated from it

Freezing Rain Supercooled droplets that freeze on contact with the surface

Gulf Stream A channel of warm water found off the East Coast of the United States

Hail Ice stones that form in thunderstorm updrafts

Humidity The amount of moisture in the air

Hurricanes Tropical Cyclones that occur in the Atlantic Basin

Jet Stream Areas of strong atmospheric current between 20,000 and 50,000 feet, depending on latitude

Lee Trough An area of low pressure that occurs on the downwind side of a mountain range

Occluded Front A front created when a cold front overtakes a warm front, cutting off the flow of warm air to the center of low pressure

Polar Vortex The circular motion of the polar jet stream spinning around the North or South Pole; it gains notoriety when it sinks far enough south to affect highly populated regions

Radar In meteorology, the graphical representation of falling precipitation created by electromagnetic waves reflecting off of falling drops

Rain Precipitation that falls as liquid, and remains so upon contact with the ground

Rainbows The visual phenomenon created when sunlight is reflected and refracted off of water vapor droplets suspended in the air

Sleet Supercooled droplets that have started to freeze into a pellet

Snow Crystallized water droplets that fall to earth as frozen precipitation

Stratus Thick, even-looking clouds that hang low in the atmosphere

Synoptic Meteorology The study of large-scale weather features, such as fronts, or high and low pressure systems

Temperature The sensation of air molecules striking the skin or a sensor

Texas Hooker A common North American weather feature that emanates from the southern Rockies, specifically, the Panhandle region of Texas and Oklahoma

Thunderstorms Strong storms marked by the presence of lightning; thunderstorms form due to strong convection

Tornado A violently rotating column of air that emerges from a wall cloud and can create extensive damage at the surface

Updraft Rising air generated by convection

Warm Front The leading edge of an advancing mass of relatively warm air

Wind The movement of air from an area of higher pressure to an area of low pressure

About the Author

Ryan Henning was born and raised in Minnesota, where he spent most of his formative years in the Minneapolis suburb of Victoria. There, he developed a fascination with the weather, thanks largely to his dad's career in the airline industry. (Ryan loved the radar!) After earning a degree in synoptic meteorology from Purdue University, Ryan worked as an aviation meteorologist for eight years. He now runs his own website and blog at Victoria-Weather.com.